SLAVE
TECHNOLOGY

Biju P R is a writer, teacher and blogger, and his research interests lie at the intersection of technology and society. He has written extensively on the interface between social media and politics in popular web journals, academic outlets and online news portals. He is preoccupied with artificial intelligence, love, sexuality and relationships and the link between technology and religion in India, in that order. Assistant professor at the Department of Political Science, Government Brennen College, Thalassery, Kerala, he is also the author of *Political Internet*, *Intimate Speakers* and *Selfie Sex*. His upcoming book, also with Rupa, is about the impacts of artificial intelligence on society.

SLAVE TECHNOLOGY

The New Age Frankenstein

BIJU PR

RUPA

Published by
Rupa Publications India Pvt. Ltd 2019
7/16, Ansari Road, Daryaganj
New Delhi 110002

Sales Centres:
Allahabad Bengaluru Chennai
Hyderabad Jaipur Kathmandu
Kolkata Mumbai

Copyright © Biju P R 2019

The views and opinions expressed in this book are the author's own and the facts are as reported by him which have been verified to the extent possible, and the publishers are not in any way liable for the same.

All rights reserved.
No part of this publication may be reproduced, transmitted, or stored in a retrieval system, in any form or by any means, electronic, mechanical, photocopying, recording or otherwise, without the prior permission of the publisher.

ISBN: 978-93-5333-566-3

First impression 2019

10 9 8 7 6 5 4 3 2 1

The moral right of the author has been asserted.

Printed at HT Media Ltd, Gr. Noida

This book is sold subject to the condition that it shall not, by way of trade or otherwise, be lent, resold, hired out, or otherwise circulated, without the publisher's prior consent, in any form of binding or cover other than that in which it is published.

For Gayu

Contents

Introduction ix

1. Bragging on the Internet 1
2. The Rise of Digital Ghettos 9
3. Gossip, Rumour and Harm to Reputation 16
4. Hate Campaign 33
5. Internet Trolling 52
6. Sexism on the Internet 71
7. Internet Narcissists 81
8. Automation and Relationships 93
9. Scam Artists 102
10. Love and Betrayal 116
11. Censorship and the Changing Personality of the Internet 124

Acknowledgements 143

Introduction

In the beginning, the Internet was configured to be a free and liberal space. It was prognosticated to give shelter to all equally. Freedom, anonymity, instant communication and a worldwide audience—the personality of the Internet was characterized by these unique traits. When the Internet started in India in 1995, optimism about its power was also huge. We believed it would disseminate unheard voices to a huge audience and reach out to the unreachable. It will empower women, and the underprivileged will find an alternative social space. It was okay to think so then. There was no Trump; no Brexit; no Cambridge Analytica; no data was leaked to third parties by Facebook; online fake news wasn't invented nor was post-truth an obsession. Now the personality of the Internet has changed. Human behaviour on the Internet is an index of the fact that our society has become predatory. A barbaric state of mind, or something worse, has been reinstated. The nature of which is yet to fully evolve; but it is a fact that not all people are comfortable on the Internet.

The Internet serves as a playing field for scam artists, cyberbullies, narcissists, haters, communalists and sexists. When you allow the Internet, which is a space with unlimited opportunity, to publish anything anonymously, to merge with the gullible mind of people, the result is the formation of a social space marked with the absence of cultured behaviour and lack of civility. Our cognitive skills such as concentration, mindfulness, memory and reflection are lost to the savageness in the digital space, where people are impatient and hooked on instant self-gratification. A heavy dependence on our smartphones, tablets and laptops has vividly changed how we talk to others, understand people and interact with them; and this digital culture is slowly eroding some of the core principles of our social world—the value system and culture. Religion is used to create superstitions and divisive social structures are reinvented. People unabashedly post abusive and foul comments that many from previous generations hesitate to utter, even in private. Natural calamities and tragedies are being treated as the new normal. People share news links, be it a stampede in a temple or a bus accident, and tag their friends, as if it was 'yet another incident'. Now, what was thought to be the real personality of the Internet is lost to a hybrid culture where emotions rule over reason.

I came to know these aspects of the Internet after a decade of being a user on various platforms. The work on this book started in 2004, which is when I began noting my observations. I have seen plenty of people who have become insensitive or more intolerant because of their excessive use of the Internet and smartphones. This book contains some of their stories.

Over the years, I have used several social media platforms such as YouTube, Facebook, Orkut, LinkedIn, Twitter, Blogger, WordPress, Viber, Skype, Instagram, petition sites, Gmail, Google+ and WhatsApp, among others, and have roamed the cyberspace as a digital ethnographer among tens of thousands of people. Many references I have cited in this book may not exist now. Some of them might have been updated, while others may have become non-functional.

I had more than one account on one social media platform.

For example, on Orkut, I had more than ten accounts at a time, which has stopped operation in India. I still maintain four Facebook accounts for my personal negotiations with cyberspace. I have closely observed plenty of social media websites and people in cyberspace. I had become an online blogger and had contributed an enormous number of articles to citizen journalist platforms, such as Merinews and India Opines, as part of my investigation. The Internet is an intimate space, which demands an out-of-the-box formula to unearth its stories.

So, what I try to portray here is not a systematic survey of cyberspace and its pathological impact on society in India, rather a first-person account of the stories of people I have known, whose lives have been affected by the Internet. Therefore, this book tells stories of the darker shades of the Internet as is seen by me, but not in statistical or quantitative measures.

To be frank, I must tell you about the lesser-known aspects of the Internet. I was naïve when I was introduced to it for the first time. Like any enthusiastic person, I too thought that the Internet would bring positive changes in our society. It has certainly brought about tremendous productive social changes in our society, but the same does not lie within the purview of my enquiry. However, I can share how my initial enthusiasm about the Internet was wiped out all of a sudden.

A few years ago, even before I was about to begin my career in teaching, I received a request from Kabsy Educational Consultancy on my personal blog. They claimed to be a global educational consultancy based out of Zambia. They said that they would be pleased if I were willing to accept their offer for appointing me as a senior academic writer. My job was to prepare project proposals for their clients across the world and offered me a whopping remuneration, which would be made on either every thirtieth or fifteenth day of the month. The payment was to be made via either PayPal or Western Union. I accepted the offer and began my temporary career. Every day, I counted the number of pages I freelanced for them. This made me happy day after day as dollars kept accumulating in my account. Every now and then, they asked me to write on a different topic. I used my Gmail account to send my works to the ID given. However, the email ID and location kept changing. Each time, emails were sent from different parts of the world and from different nationalities. When I asked them about the changing locations, they told me that they were a global firm. I kept a close watch on their websites and found nothing unusual. I also checked blogger.com, verified their Facebook updates, and followed their YouTube videos. Their website and other social media profiles requested clients to remit fee online. Their website www.kabsyconsultancyservices.org, which no longer exists, said if someone wanted to know more about them, they could drop a line to the ID provided. They were actively present on all social media platforms.

On the fifteenth day of my job, there was no communication about the payment. I had already sent them the soft copies of my educational certificates, PAN card and other identity proofs. I am not sure whether they were misused, and I was afraid that my PAN would be used for illegal activities. When I did not hear from them after several emails, I decided to publish all the articles on my personal blog. I never heard from them again.

It is that easy to be cheated on the Internet; one can see its personality changing fast. Such vulnerabilities that arise from the benefits of the Internet spoil one's life on the social media landscape. Relationships get affected and there has been an increase in cybercrimes due to the anonymity the Internet provides its users. This book does not discuss the benefits that we derive from an emerging techno-social culture. It deals with the dangers and threats that I have identified, which are hidden behind the hype of the positive aspects of the Internet. So, when we acknowledge the positive changes in the Internet and digital world, we fail to recognize the perils it contains. There is no doubt that the Internet makes our lives easier, but its hazards remain hidden behind these comforts and are unknown to us. In my social media encounters, such as the one with Kabsy, it is evident how we have become more vulnerable to the ills of social media. It hurts people and makes them helpless, sad and lonely.

The Internet gives you opportunities to connect instantly, it gives you a wider audience to build your network and collaborate with them. Some would see these as the positive attributes of social media, which would bring about big social changes. They call it the 'people power';[1] others call it 'mass power'.[2] However, one would also see that, as in my case, social media caters to our vulnerabilities, and I had to adjust to this reality. I knew that I was not the only one who had been duped, even though I hadn't shared the incident with anyone. However, when I decided to write this book, I made up my mind to include my experience, which could help substantiate the fact that the Internet is changing from its initial enthusiasm to a hostile ecosystem. If educated people easily fall prey to such devices, the way in which cyberspace is used to victimize ordinary people is quite distressing.

My understanding about social media platforms is that it is more problematic in Indian life. *Slave Technology* is about my anxieties about the Internet and digital technologies and about the untold stories of India on social media. One may notice that there is little attention of the media on this topic, fewer discussions even among the learned professional community about this invisible India I am about to discuss.

This is also a book about 'hate culture' and other divisive

tendencies on social media, and a personal journey into the so-called hype around the 'people power' of social media. I have used the term 'slave technology' to signify the consequences of the Internet in the lives of ordinary people, who look at this with greater hopes and dreams. These hopes and dreams are fallacious claims on social media and its 'mass power'—that it can bring significant changes in society, old hierarchies can be democratized, social divides can be brought down, it can empower people on the societal margins, organize public scrutiny of the conduct of public services and reinvigorate civil society.

Over the years, there has been a phenomenal increase in the number of people boasting in social media too. In academic venues, scholars started to make positive correlation between democracy and social media. Television debates air prophecies about social media. The print media carries full-page features that boast social media as making bigger changes in the lives of the ordinary people.

Meanwhile, there are many ordinary individuals whose experiences were opposite to what were the commonplace assumptions about social media. It is ruining our social skills and corrupting the moral foundation of society. Everything that matters to us, be it relationships, emotions, attention spans, values, language and culture, that have shaped our social world in the past are affected by social media.

This book scrutinizes whether such conjectures are illusory and investigates the impacts of social media on society. It deals with the ways it ushers unintended consequences; how divisions on the basis of caste, class, religion and gender are reinstated on digital platforms; how it serves as a platform for trolls and spoils someone's reputation within seconds; and the way in which paid social media campaigns fabricate facts. In short, it elucidates the manner in which social media is used by the elite to control the society and our minds as they unite to oppress and divide the masses, to consolidate their cultural and socio-economic power in society.

People who are born in the pre-Internet era can easily identify how the onset of Internet in India changed the social landscape all of a sudden and what prompted people to migrate onto it; the contours of these changes are more apparent to them. As I am nearing middle age, and with my increased involvement in this field of research, I have realized the increase in cybercrimes.

While I was looking for anything fitting my concern, there was no information depicting the darker shades of social media barring one or two. I became more and more obsessed with this side, which encouraged me to write this book. Besides, many of my friends and peers also share the Internet phobia.

I have used many neologisms in the book that depict the ill-effects of social media, which systematically substantiate my core assumption in the chapters that the Internet is changing into a hostile ecosystem in the sense that it reproduces old social divides in new forms.

While there are many books on the applications of social media platforms, there is little on the harmful side of social media. This book focuses neither on the statistics of social media nor on what social media could do for us. It is just an intimate conversation about the social media experiences I have come across and what other people, whom I have met on the Internet, have told me about their experiences. Some of the names of people in this book have been changed to protect their identity.

Endnotes

1. *The Wisdom of Crowds: Why the Many Are Smarter Than the Few and How Collective Wisdom Shapes Business, Economies, Societies and Nations* by James Surowiecki (2004) explains that the crowd is smarter than the elite few.
2. Chris Anderson, in *The Long Tail: Why the Future of Business is Selling Less of More* (2006). Glenn Reynolds, *An Army of Davids: How Markets and Technology Empower Ordinary People to Beat Big Media, Big Government, and Other Goliaths* (2006) speaks about the great mass advantages of technology.

1

Bragging on the Internet

I used to talk to strangers at the airport and railway station. I met them for just a few minutes and spoke to them about politics, sports and films, and we exchanged useful information on the same. But now, these people have developed 'text neck' as they bend their necks to look at their smartphones.

—Shashank Gupta

I have a lot of time to myself as I have uninstalled WhatsApp. Now, I read the newspaper. I can feel the joy of mornings. I have reclaimed my lost reading habit. I have many things to think about now.

—Shyamla Kumar

These are a few replies I have collected from people who have uninstalled WhatsApp.

The other day, I was talking to a friend and colleague from work who had quit a WhatsApp group consisting of his colleagues. While WhatsApp groups and other digital platforms help us connect with each other, my friend told me about the meaningless conversations that took place in the group. A member posted a picture with the caption, 'Hi friends! I bought a new car—Toyota Altis. It's a dream come true. It's a good car.' All of a sudden, a comment followed appreciating the member for his new acquisition. Comments came flowing in; some even extending to the length of a few scrolls! Most of the comments compared it with other cars and its merits and defects. But none of the people in the group were automobile experts nor did they own a car! Somebody started saying their friends had the same car, but they had sold it at a revalue shop as they did not like its performance. A lot of human energy is being poured on this exercise. The discussion continued for a few days until another member of the group started a new topic. As their attention moved on to the new topic, they created a huge stockpile of mindless conversations over the next couple of days. Their involvement in the discussion gave an impression that they had solutions to serious problems of the nation! Politics, economy, culture, movies, art, gossip—almost everything found their interest. My friend said there were people on the WhatsApp group who boasted blatantly. It goes something like this: 'I have hundreds of citations on Google Scholar. I am one of the most-cited academicians in the field.' The other day, somebody said, 'I have published a number of articles and presented over a hundred papers.' The posts didn't offer anything substantive or insightful. These posts can be simply interpreted as: 'Respect me. I am awesome. My works are more recognized than others in the same profession.'

You might have lost a good night's sleep over continuous WhatsApp updates. People spend their nights chatting in these groups as if they have nothing better to do! You go to the group only to realize that it is full of messages that do not interest you:

videos, photos, jokes, motivational messages and advertisements. Just name anything in the world; you have all that in the group! As if the meaningless content was not enough, an array of smileys followed them. After a few 'His', 'How are yous?' and 'What elses?', there would be dead silence for a while in these groups. Some people, in the meanwhile, take the initiatives and post something in the group. What follows is a smiley war among the members of the group. As the group grows older and the members have gossiped about almost everything under the sun, the group has just two kinds of messages: good morning and good night! This is the time when you feel the pointlessness of being a part of this group. All you can think of is, 'When can I get myself out of this!'

Somebody creates a group or adds you to one, without notifying you. It starts innocently, but you soon realize that it has affected your peace of mind. Every time you engage yourself in something productive, such as reading a book or listening to music, a WhatsApp notification disrupts your focus. As you mute the group, it still pops up on the mobile screen and flickers until you check the notification. You start anticipating messages and feel the need to respond to every question asked in the group. By the time you get back to what you were doing, you realize that you have wasted a good amount of time looking at a picture or an advertisement or listening to an amateurish political commentary that was shared on the group. Perhaps these updates did not even require your attention as the content circulated is not even authentic.

When you put up a WhatsApp status, you frequently check the number of viewers and those who have seen or ignored it. In this way, you create your social world—those who view your status are your friends, and others become your enemies. This was what Fidha, one of the teenagers I spoke to during my research of this book, told me. She was obsessed with the status she put up. Those who ignored her could not be her friends, and she would create her social connections on the basis of those who saw her status. This is bizarre, but such is the new world of connection.

A friend of mine who is a banker told me that although he wanted to quit the groups on WhatsApp, he could not, for the fear of personal conflicts. Another friend said that he had to uninstall the app as acknowledging every update and the unnecessary enquiries when he didn't reply robbed him of his time. This can also be psychologically tormenting and creates a complete mess of things. You can find similar groups in organizations or social spaces such as schools, government departments, public sector undertakings, private offices, corporate houses, neighbourhood, alumni groups, and so on. Such groups can also be spotted on Facebook, Twitter, Instagram, blogs, etc.

The basis of its formation is unique to India's social structures—caste, religion, class, language and hate. What motivates people to become a part of such groups is also a divisive technique. From WhatsApp to Facebook and others, how people are stupefied by pettiness is beyond explanation. But there are other distasteful conducts in cyberspace.

Radhika was a lesser-known woman activist before she took to one of the social media platforms for reaching out to a wider audience, and incidentally, she became popular on Facebook. Perhaps it is the desire to connect beyond her physical limitations that prompted Radhika to write on her Facebook timeline. Day after day, she posted opinions about almost everything on it. Her opinions ranged from the conversations between the US President Donald Trump and the Supreme Leader of North Korea Kim Jong-un to the Chang'e Project, which is an ongoing series of robotic Moon mission by China National Space Administration

(CNSA). Gradually, people started responding to her Facebook commentaries, including her profile picture updates. I noticed that most of her followers were men. With time, she got invited to literary events, seminars, discussion forums and other public venues. However, nobody seemed to question whether it was her merit or the social media presence that made her popular. This, sadly, has become the rule now, not an exception.

Another profile that I observed was that of Parvathy Menon, who is an assistant professor at a south Indian college. The profile was created in August 2018 and her friends and followers had spiralled to 5k and 1k, respectively, by December 2018 when I accessed her profile. Her profile description says that she is a 'poet, social critic and classical dancer'! Her profile picture, updated on 5 February 2019, had 1.2k likes and 154 comments within just three days of updating it. Almost all the profiles on her friend list were of men. On the other hand, meet Sam Taj Kannur, an assistant professor of English Literature, who has close to 5k friends on Facebook and close to 1k followers on Instagram, as accessed in December 2018, just like Parvathy Menon. He had shared a TikTok lip-syncing video on Instagram and Facebook. In the video, he expressed his feelings through random, amateurish dance moves. However, none of his videos gets more than fifty views. This behaviour has now become a new normal, and a new culture is being formed, which is widely circulated on social media, aiming particularly young people. The updates from the people belonging to this culture appear to be unintelligent, silly and have nothing to do with the serious problems that the world is confronting. However, when women like Parvathy Menon put up posts, even though they are silly and meaningless, they grab more than 2k likes and over 200 comments within a few days. May my personal opinion be apparent in seeing their updates as silly and meaningless, but this aspect of cyber culture made me really curious. As I looked on, I found that most of the followers on the female profiles were men. Had they been profiles of men, like Sam Taj Kannur, the viewership would have been different. What works here is simply male libido. There is little possibility that the engagement on such profiles is with an intention to have a serious dialogue.

When I was younger, hard work was acknowledged with probably a bouquet of flowers and a basic standard was set as a vanguard of merit. Now, a simple photo gets more than a thousand comments. This seems to be a waste of time and spurious creativity. Besides, it hampers social skills and encourages negative emotions such as jealousy, rivalry and narcissism.

People share positive events and achievements from their life on the Internet. They share posts about their promotions, engagements, newborns, etc., with their relatives or even acquaintances. In doing so, they actually spread positivity and happiness. It is good to encourage their efforts and be a part of their accomplishments. However, there is a very thin line between seeking appreciation about the highs in one's life and bragging about them. Bragging incites jealousy, a feeling of both inferiority and superiority, makes others envious of one's successes and vis-à-vis; this spreads negativity among those who are unable to accomplish such things in their lifetime. People use false modesty as a context to tell others how great they are. Or, their posts fail to provide any informative content to their audience. When bragging, the information one shares and the people they share it with, both matter.

Humblebrag, a term coined by comedian, writer and producer Harris Wittels, refers to posts on Facebook or Twitter that tell the world how great someone is. However, it is downplayed under the guise of modesty. Consider this: 'While I was driving my new Range Rover Evoque, I found a broken-down school bus. Picked up a few school students and dropped them at their respective stops.'

People brag because they want to be famous and valued. They want their followers to tell them how impressed they are with their accomplishments. And, when the real world fails to do so, some people take to the social media to feel more worthy and be praised.

Bragging is also a tricky business on the Internet. We can see how people react to boasting in the real world. On certain social media platforms, there is no face-to-face interaction, and people miss social cues during communications. People usually seem uninterested, frown and use gestures of disapproval, which tells the other person to modify their behaviour accordingly. To overcome these, one may, consciously or subconsciously, try to veil their egocentric, narcissistic and self-gratifying image by either being modest or refusing to respond.

Posting on social media is a way to let people into one's life. At a basic level, WhatsApp, Twitter, Facebook, etc., are replete with people who are comfortable not only announcing their accomplishments but also sharing every minute detail of their day.'

However, this type of bragging presents only one side of the story. Those who brag about their achievements indeed do it while hiding the unhappy side of their lives. So, when somebody says that the Internet is the future of democracy, I wonder why they don't address the aspects of its darker side that exists. Political communication and other activities on the Internet have already fallen prey to many divisive forces.

By the time I finish writing this book, I feel that there will be fewer people who would still believe in talking face-to-face, which helps understanding emotions better. It is basically a valuable human quality. They are but an old tribe in the digital age, where people feel stressed when they don't get enough 'likes' for their pictures on Facebook. Harper Lee took fifty-five years to write *Go Set a Watchman*—the sequel of *To Kill a Mockingbird*. During this time, she remained silent. It took almost twenty years for Arundhati Roy, who won the Booker Prize for her debut novel *The God of Small Things*, to write her second novel *The Ministry of Utmost Happiness*. It may seem unbelievable, but it is true. Not bragging about yourself will get you fewer likes and replies on your updates. In times where every aspect of life is automated and mediated by technology, one is forced to migrate to this world. Otherwise, one doesn't even exist!

2

The Rise of Digital Ghettos

A family of a husband, his wife and their two children are seated at a table in a restaurant on a busy street. The father swipes the notifications on his smartphone, and his wife opens the Facebook app. The elder child is closing in on the high score set by his/her friend on one of the trending games; he is on his tablet, while the youngest one eagerly looks at them, expecting to get his/her smartphone very soon.

At a grocery shop, a teenage girl, while looking at the items on the shelves, gets a WhatsApp call from a friend. She takes the call even though it is not urgent.

At a busy fabric shop in the town, boys are seen taking selfies and, of course, not engaging with the purpose of their visit to the shop.

These are scenarios where you notice someone or the other is hooked on to their touchscreens. At places such as railway stations, shopping malls, weddings, get-togethers, etc., it is easy to find people glued to their screens. They fail to realize that they are missing out on interactions that involve face-to-face communications, which societies have cherished for centuries.

Even during a family outing, technology engages people into petty photo updates; they end up not talking to each other and miss enjoying the moment. They have shorter attention span and they are losing the ability to pay attention to details. In fact, they have become the slaves of technology. The incidents from the restaurant, grocery or fabric shop may seem fictitious, but they are neither stray examples nor made-up cases of people distracted by technology or those of depleting human sensibilities. Classrooms, family gatherings, conference halls, peer circles, offices, outings with friends, talks over coffee and similar human behaviour have been affected by the advent of the Internet.

Their social world is made of activities in the digital space. Their thoughts are being automated. Look around; people are vulnerable to the nauseous content on the Internet, comprising misogynistic and hateful comments, paid campaigns, brand-marketing strategies, tainted reputations and other crimes.

These days, people who take pleasure in sharing meaningless content throng the Internet. They also engage in scanning updates without any purpose. Their behaviour has changed the personality of the Internet. Due to the infinite possibilities of being able to upload numerous pointless things, the quality is lost to quantity and professionalism is lost to amateurism. This changing personality of the Internet has negatively affected human skills and the moral foundation of the society.

What I have been able to identify here is not a nationwide report of the impact of the Internet on human cognitive skills, but an intimate portrayal of people and their behaviour in cyberspace for more than a decade.

Are We Connected?

The smiley outbreak has influenced emotional intelligence. Some people are too fond of using emojis on the Internet and they have contaminated their minds. Any number of meritless smileys being sent to any number of meaningless updates on the Internet makes it a trivial space, where emotional skills are lost to a new form of interpersonal communication. Indeed, the smiley epidemic has spoiled the personality of the social space, thereby dehumanizing emotions of human beings. Values and emotions have been smashed into shards. Emoticons share only one side of the story. We are forgetting that a smiley, or any emoticon for that matter, makes us that distant from the real person.

Diminishing Linguistic and Reading Skills

The Internet is a control-free medium of information-sharing. Certainly, the language that people use to convey their message is experimental and ever-changing. In 1994, when the Internet was still text-oriented and dial-up friendly, and email was just beginning to find traction into daily life, an American essayist and literary critic, Sven Birkerts, wrote the book *The Gutenberg Elegies: The Fate of Reading in an Electronic Age*. He predicted that flooded by electronic devices and data into almost all aspects of life, what most of us risk losing are what books and reading culture have provided—the occasions to read, think, pause, reflect, revisit, memorize and reflect some more things.[1] Digital technology has given birth to a type of electronic language that carries meaning only when used in cyberspace and in a specific situation meant for it. Internet slangs are used without considering the context. Linguists refer to this communication as 'texting', while others prefer the term 'textese'. Texting caters to the short attention span of people in which they are obsessed with instant responses. People desire for adjusting everything in compact forms, which also contains features to make them more expressive. For example, 'LOL' is an Internet slang for 'laugh out loud'. No one uses LOL in face-to-face conversations, but on cyberspace, it is frequently used. The point is that texting is in no way similar to written and spoken language because it has developed only in cyberspace. It is debatable whether it carries any meaning outside cyberspace. An increasing use of acronyms, otherwise called Internet slangs, such as 2N8, CU, GN, GR8, MSG, damages our linguistic competence and is unparalleled in the recent history of our civilization. Of course, we all use acronyms, but the language developed in cyberspace carries meaning only in specific situations. Outside those, the particular acronym carries no sense at all. That said, the Internet has certainly enriched language. However, there is another side to this as well. Many have also stopped reading books completely as they find readymade information easily available on the Internet. This is likely to rob us of an eye for detail, reading skills, and thereby, intelligence. In the age of short attention spans, libraries and bookshelves have turned into scarecrows. Although it is likely that we read more, this is at the cost of quality. It has certainly affected our cognitive abilities.

Bad Taste and Mob Rule

On the Internet, everyone is a creator. Some become famous in a blink of an eye. Amateurism thrives to negate professionalism, moral standards and culture.[2] The silliest of things become market hits. Music is withering away to amateur garage bands. Movies and television programmes are switching to the bandwagon of amateur YouTubers. Author and entrepreneur Andrew Keen has rightly said that, 'When ignorance meets egoism, bad taste meets mob rule.' The most bizarre thing about many Internet celebrities is that none of their work merits any attention, nor do they pass the test of professionalism. Ironically, what gets them fame can also get disrepute.

Normalizing Violence

Violence pervades our society and it is not surprising that it now invades social media. Some people regularly watch sportspersons beat each other on the playground; mob lynching of Dalit men and women, stripping, mass shooting and violence against women. There are lots of people on social media who are exposed to violence as a routine. There are reports of online criminal behaviours such as bullying, stalking, harassment and invasion of privacy. When we can block violent content, it (think of a friend or a website that shares violent content) means that we are normalizing what otherwise would be unacceptable.

Besides, cybercrime also includes using social networking sites to cheat people of their money and/or affection, or to recruit youth into illegal activities. Usually, they befriend you and offer a sympathetic ear, thereby gaining your trust, and then strike. The criminals also use social media to perpetuate crime and to hide from law enforcement agencies.

For any silly reason, citizens fight on social networking sites, and fall into trouble. For instance, in 2014, a Chennai-based businessman was arrested for allegedly posting abusive comments about a lawyer on a social networking site. The feud was related to a dog show and the subsequent announcement of prizes.

While these types of crimes are being normalized on the Internet, it would invite either punitive action or incite a reactionary mob offline.

Social Media Addiction

Some people use social media compulsively and excessively. Their world comprises the social networking websites such as Facebook, Twitter, Instagram and Snapchat, and they don't even bother when the use of those platforms takes over their lives and they have a negative effect on real life and relationships. They are not even able to sense that it robs them of their life, time, energy and peace of mind.

They check their phones constantly for Facebook or Twitter notifications. They become nervous when nobody reacts to their update instantly. They feel low when they do not have access to social media.

The false sense of connection that comes as a by-product of social media is so dangerous to our sense of self. Repeated exposure to social media sites leave users with an 'identity crisis' and in a state of emotional comatose. Cyberspace has created a generation, which is compulsively looking for likes, shares, comments and recognition of petty and silly updates that would have nothing to do with ground realities of life.

Our vital human skills are spoiled by the emerging digital culture. Cognitive abilities are automated and narcissism is celebrated. Selfies show the magnitude of self-love. Digital culture has maligned our skills, including reading, attention, expression of feelings and linguistic skills. Handwritten letters, spending time with people without getting distracted by phones, writing without making spelling errors, talking to family members, boosting one's word power without the help of a dictionary and driving on unfamiliar roads without a navigation system prompting directions seem to be things in the past. This scenario, where the Internet is turning people into slaves and taking over the cognitive human abilities will continue in future. It is only when we realize that we cannot move further that we will respond to it. Otherwise, the meaning attached to relationships and being human will be conquered by the Internet.

Endnotes

1. Sven Birkerts, *The Gutenberg Elegies: The Fate of Reading in an Electronic Age*, Boston: Faber and Faber, 1994.
2. Andrew Keen, *The Cult of the Amateur, How Today's Internet is Killing Our Culture and Assaulting Our Economy*, London: Nicholas Brealey Publishing Ltd., 2008.

3

Gossip, Rumour and Harm to Reputation

Over the years, I have realized that I am not alone in facing cyberstalking and online attacks. Many people face it in the form of trolling, stalking and violations of privacy.

The other day, I received a scathing comment on Facebook on one of my articles I had shared. The commenter is a friend from college, who is now an advocate at the Supreme Court. The article is a cautionary tale critical of the ruling government. 'I read the entire thing. I do not want to get into any verbal duel with you on what Modi Sarkar is all about, your perception, your thoughts or your Facebook page. I have no business to interfere, but what struck me is the use of certain terms,' he said. The comment was a disconcerting one. He continued to vent his distaste over the article to such an extent that I began to think our friendship was on the brink, because I criticized the Modi government. Anyhow, our arguments continued for several days, until one day I stopped responding to them.

In 2014, Maria Sharapova, the Russian tennis player, faced flak on the Internet when she told a journalist that she was only able to identify the soccer player David Beckham and the basketball legend Michael Jordan among the VIPs in the Royal Box at Wimbledon. The venue also had Sachin Tendulkar, the Indian cricket legend, who watched the match that day. Irate Indians, many among them from Kerala, were infuriated by her inability to recognize the cricket legend. They lashed out on her Facebook page in extremely insensitive and crass language in their local tongue.

In 2015, when Pamela Denise Anderson, the Canadian-American actress and the brand ambassador of People for the Ethical Treatment of Animals (PETA), wrote to the then Kerala Chief Minister Oommen Chandy, suggesting that the elephants used for the famous Thrissur Pooram be replaced by toy elephants, all hell broke loose. This annual festival of Kerala has a grand procession of elephants as a part of the rituals. In no time, her Facebook page was flooded with abusive comments by fanatics.

These incidents are random. However, all of them signify the direction in which the digital space is moving towards. What do these incidents indicate?

A common trend can be traced based on such incidents. Some people operate with stakes. They have a specific agenda for spreading emotionally charged and distorted information. It is the outcome of a well-though-out plan of action; spoil the reputation of some people; cause dissension in society; and spread violence everywhere. It is a fact that both gossips and rumours are establishing an official version as final and unquestionable.

Mob Lynching and the Internet

In a string of events in recent years, many innocent people were lynched by unidentifiable mobs based on false information spread on the Internet. Tripura, Uttar Pradesh, Rajasthan, Bihar and Jharkhand had witnessed a spate of violence persuaded by fake news on the Internet, particularly via WhatsApp.

Some instances of fake news from the recent past are:

1. The newly released Indian currency notes of 2,000 have GPS tracking chips embedded in them that could locate bank notes hidden several feet under the ground.
2. The rumour about the shortage of salt made consumers flock to market salts in some Indian states.
3. In southern India, the rumour about measles and rubella vaccine thwarted a government immunization drive.
4. The rumour about child abductors triggered mob lynching of innocent people.
5. Rumours about an occult gang chopping off women's braids in north India spread panic, and consequently a woman, belonging to a lower caste, was killed.

Some fake stories perpetuate from India's rising religious and caste conflicts. For instance, images purportedly showing attacks against Hindus by the Rohingya Muslims in Burma circulated on social media has given rise to hatred against Rohingya Muslims in India.[1]

Another instance of fake stories was a video showing two people being beheaded. The text in it said that they were Indian soldiers killed in Pakistan. However, it was actually taken from a footage of a gang war in Brazil.[2]

The very underbelly of rumour is that it is used as a weapon to attain political goals. It emotionally appeals to the masses even if it is illogical. Usually, it spreads by word of mouth. Every generation has its own rumour-minting tools. In the absence of mobile phones, television and the Internet, information, often fake and exaggerated, took longer to reach people and spread slowly. The communication channels were person-to-person.

Rumours were exchanged from one community to another and from one town to another.

Now, there are new mediums for spreading fake information—social media. It spreads among a wider audience within the blink of an eye. The number of lynching cases triggered by rumours spread on social media in India is alarming. For example, *The Quint* reported that 90 people have been killed in mob violence since 2015.[3] Between 2012 and 2019, 29 persons have been killed in cow-related hate violence, 25 of whom were Muslims, says *IndiaSpend*, a data journalism website.[4] The year 2017 has been the worst for cow-related murders, killing 11 people and cow terror attacks involve mob lynching, attacks by vigilantes, murder and attempt to murder, harassment, assault and gang-rape.[5]

Although many lives have succumbed to mob lynching spurred by rumours, the statistics vary from one source to another.

According to an *India Today* report, social media rumours about child lifting and cow slaughter created 16 cases of lynching in two months—May and June 2018—and it resulted in 22 deaths, including that of a transgender's.[6] In an analysis of more than 30 million English news articles (print and online) published between February 2014 and July 2018, 31 people have been killed in two years due to social media rumours. This includes a fake WhatsApp video of child lifting, which was actually part of a safety video produced by a child welfare group in Pakistan.[7] Between 1 January 2017 and 5 July 2018, 33 persons have been killed, and at least 99 injured in 69 reported cases of mob lynching, reports *IndiaSpend*.[8]

The creation of lynch mobs is not well researched. Lynch mobs also exist on social media like a swarm of bees. Throughout history and in epics, rumours have been used to mobilize people. In the US, for example, the white Americans used mob lynching as a means of terrorizing the blacks. In India, a dominant community uses lynching against the minorities, such as lower castes and Muslims.

Lynching also involves setting victims on fire or hanging them from trees. The purported aim of this sort of brute behaviour is to reclaim the supremacy of one community over the other. Lynching terrorizes the 'enemy' and sends a chilling warning to those who stand against anything in the interest of the dominant group.

India is undergoing a phase of intolerance against the 'other'. There has been a huge increase in cases of mob violence in the country in recent years. The phenomenon of rumour threatens India's democracy and the freedom of people. However, mob lynching in the age of Internet is a bit more complicated. For example, as many as 33 people have been killed and 99 have been injured between January 2017 and July 2018, says a report.[9] The year 2017 remains to be the worst in reference to mob violence, killing 11 people.[10] This data alone shows the surge in mob violence in India. These include cases of a thirty-two-year-old man who fell victim to rumours about child lifters in Bidar of Karnataka, the lynching of two men in Karbi Anglong of Assam, and the death of three in Tripura in June 2018.[11] The fact is that this type of mob lynching was spurred principally by rumours on either WhatsApp or other social media platforms. According to an *IndiaSpends* analysis, between February 2018 and March 2019, 78 incidents were reported, and the principal source of these rumours was social media and WhatsApp, which resulted in the steep rise of mob violence. The year 2018 had seen almost all the violence spurred chiefly by rumours on social media.[12]

Clearly, mob frenzy builds up when certain emotional issues are stirred. The news spread is just rumour. No such issue (in most cases) was ever actually reported; and the perpetrators of these crimes continue to be unknown. Mob lynching triggered by rumours on the Internet is only the tip of the iceberg. There is more to such crimes being perpetrated on the Internet. That said, what motivates this type of brute crime is the social structure unique to the society in India. Therefore, naturally, when rumours spread about a man, belonging to a lower caste, marrying a woman of an upper caste, or upper-caste women talking to strangers, the upper-caste patriarchal mindset acts in order to protect the 'honour' of their community. There is an element of emotional appeal behind such rumours on the Internet. As long as we have our feudal structures intact, social media rumours will continue unabated and the crimes will also increase in the future.

We have to get away from these types of crimes unique to India. One way to do so is to sensitize children to this menace, preferably in schools. School curriculum ignores the idea of sensitizing the young generation about the evils of fake news on social media. Even informed and educated people are unable to distinguish between good and bad news on social media. Therefore, there is no point in blaming children falling prey to social media rumours and perpetrating more crimes, resulting in collateral damage.

Gossip on the Internet

Some people share pictures of almost everything they have captured on their camera. They comment, update and like almost everything. Some profusely articulate their views on politics, class, race, gender, religion, sexuality, community, etc. Others create social media pages and forums to get people to join in debates. In fact, their behaviour often brings about a backlash. The shared comments, photos, etc., are often discriminatory, and at times, a full-blown offensive on the people who may be powerless.

The very nature of social media is that anyone can create a group that could initiate gossip against certain religions, linguistic minorities, people with different features, sexual minorities, differently abled people, etc. Certainly, social media platforms have become a place for gossipmongers, and a corner to cultivate hatred for others.

Differentiating between real and fake information on social media is becoming increasingly difficult. Social media platforms such as Twitter, Facebook and WhatsApp allow anyone to post their opinions and their stories for the world to see, no matter what.

One way to prevent the spreading of fake news is to create awareness about crosschecking the source of the content before sharing it further on the Internet. As information is shared quickly, it is becoming more and more difficult to identify the source of the stories. Obviously, assessing the accuracy of the information is difficult.

Based on a Harvard study,[13] NPR, a non-profit media organization, has corroborated that students have poor ability to differentiate fake news from real ones.[14] The problem is severe in India. The impact it has on the minds of the younger generation is more worrisome as they are at an impressionable age. Since there is no gate-keeping mechanism on social media, any activity out there is left to self-reflection and personal judgement. Hence, it is also very easy to mislead, misinform and control people's thinking.

On August 2018, a report[15] published by *Storyful*, a social media intelligence and news agency, with the title *India and Misinformation*, found that WhatsApp is the most prominent platform for spreading misinformation; prominent instances of misinformation are linked to political figures and parties, whereas when it comes to religion, particularly Islam, Islamophobia is a common theme.

However, there is more to this gossip on social media. A wave of social media content and WhatsApp messages arrive on almost every Indian's smartphone every morning. They vary from the simple 'good mornings' and 'good nights' to the more optimistic revelations of anti-aging drugs invented. In between, there is everything: sinister warnings about smugglers operating in the area, child lifters, inter-faith marriages, movie gossips, pornography, bloopers and gossips about celebrities and so on.

We are increasingly becoming interested in things that we otherwise would not have been interested in, had it not been for these social media platforms. We talk more about other people's private lives, judge them and find pleasure in silly and meaningless things. Social media has enabled this and has influenced the frequency of such things taking place.

Gossips, like viruses, spread easily. Those on celebrities are more common than other kinds. On the other hand, gossips about yoga guru Baba Ramdev and politicians such as PM Narendra Modi, Rahul Gandhi and Manmohan Singh do not spread as easily. Writers, thinkers and sports icons are no exceptions. Movie stars, cricketers, Internet influencers, celebrities, TV anchors and other public figures have become victims of this gossip culture. Young people easily accept what is spread on the social media platforms as truth, not realizing that they are rumours. The younger generation easily believes in whatever is popular. They do not check the sources, credibility and authenticity of the information disseminated.

The change in the lifestyle of the growing population with smartphones has increased the number of Internet users. A moral panic is created and the Internet and social media pose a risk to teenagers and this becomes actual issues in the society. Cyberbullying, name-calling, gossiping, making threats, spreading rumours and sending malicious messages through emails have augmented the problem. The youth has greater social identification through social media. It affects the character formation, personality and moral development of children. Chat forums and user profiles allow the young and the amateur to access unlimited and uncensored information from their peer groups around the world. However, they do not understand that there is a difference between facts and rumours. Social media platforms are unleashing monsters with no perspectives. A majority of the heresy is developed through social networking sites. It has become a paradise for gossip, which infests people with innuendos, insults and half-baked truths.

Harm to Reputation

Harm to Reputation is a phrase used on the Internet that means tarnishing the image of people who have good public visibility, built by honest means over the years. Unlike defamation, which is a crime according to the law in India and a brute intrusion into a person's privacy, harm to reputation is a new culture being developed in cyberspace as an organized attempt to paint a person in bad light. But a narrative is built to show that the person had always been this way and they have evidence to prove their point. On the basis of forged evidence, a person and his life is brought under public scrutiny. It happens on every platform—from WhatsApp groups to Instagram. Everywhere, people thus tend to hear the same. Nehru and Gandhi are discredited. Women are grilled for the opinions they have. Politicians, writers, intellectuals and others are subjected to this sort of vicious harm to reputation. It is indeed a crime, which has not been recognized and labelled under any law in India.

Gossip and rumour have been prevalent in our society. It is everywhere; nothing remains out of the radar of those who want to create some sort of gossip. Be it celebrities, friends, neighbours, teachers, court cases, youth icons or high school drama, gossiping is common among everyone. It has become a social behaviour, even though it is not problematic in certain cases. However, as the saying goes, too much of anything is bad. The fake news takes malicious turns as people get hurt and their reputations are severely damaged.

Some people log on to the Internet, spread gossip, mobilize people on the basis of rumours, damage the reputation of good people by spreading lies, distorting facts and perpetuate violence in the society. They use the Internet as a weapon. It has given birth to mobocracy. Their number is increasing day by day, and there is no sense of fear among the people indulging in defamatory activities. Many innocent people have fallen prey to mobocracy. It is a rule of a dominant group of people, unrestrained by any principles.

There are many instances in which people cooperate online to expose and punish what is perceived as repugnant behaviour. People reach a consensus to punish somebody based on an agreed sense of right behaviour.

When a certain kind of people spot a picture or a video of a woman drinking alcohol, they begin to teach her the principles of being an 'ideal woman'; and the same goes with smoking. Her name, office address, picture, home address and other personal details are publicized online.

If somebody has refused to pick up their dog's excrement on the subway, they are subjected to some sort of moral preaching about right conduct. Even though this may be considered an attempt to keep the surroundings clean, it does not allow someone to bully the other person to abide by the rules. People simply troll others for human errors or faults that are not intentional. Social media has opened an avenue where people simply spread rumours and gossip about others. How far the sanction can go and what the extent of the unwarranted shame, fear and economic harm caused by it is, is complicated. There is just a thin line between what is right and wrong, and it is often confusing for people to judge. Social media is a place where a group of people can go to any extent to harm someone's reputation. Different types of defamation cases come up in cyberspace. Slandering, discredit, tainting, smearing, disgrace—the possibilities are numerous.

Although the Internet has been a source of empowerment, it has also been a platform of significant harms caused to oneself. Even when wrongdoings deserve sanction, a mob's collective punishment may lack accountability and be disproportionate.

Why Rumour and Gossip

There is a hidden agenda behind the rumour and gossip being spread—destroy your enemy by causing mental or maybe even physically harm. So, in the end, casteism destroys the lower castes, misogynists unleash sexist attacks on women, dominant religions threaten those in minority, and power hungry people destroy their competent rivals for more power in society. The creation of gossip and rumour is very clear. It spreads dissension in society, terrorizes people and finds the dominant narratives as ruling ideas.

Although most people gossip to either strengthen their social status or out of boredom, the gossip and rumour spread across the social media landscape in India caters to a particular kind of greedy audience. It speaks to a particular kind of ideology, a certain type of social system and tries to win a particular belief system over another. It tells us about the country's peculiar social structure. It appeals to caste, religion, language, class, gender, etc.

Those who wish to leverage factors unique to India's social structure plainly use it against the other groups. So, gossips and rumours instigate a different kind of mindset—communal, caste-based, religious, regional and other divisive tendencies.

Politics, the Internet and Rumours

The circulation of unverified information is an important part of political communication. Rumour plays a significant role in shaping public perception and bringing electoral margins to either the Right or the Left of the political structure. For example, information, which was considered pro-right wing in India, was rarely accepted by the common man a few decades ago, probably because of the strict mechanism where unverified and unauthentic information did not get out easily. Today, social media has given a huge visibility to the right wing and their subjects. Heresy such as the Taj Mahal was once a Hindu temple, Bollywood has more Muslim actors, luring Hindu women into love jihad, etc., are spreading rapidly.

Here are a few more examples of 'facts' spread on the Internet:

1. Science and ancient India:[16] Indians invented the Internet, satellite technology,[17] plastic surgery, genetics,[18] the airplane[19], inextinguishable fire, nuclear arsenal[20] and advanced stem cell research.[21]
2. Darwin was wrong.[22]
3. Dung and urine from cows can cure deadly diseases such as cancer.[23]
4. The ritual of burning ghee and other food stuff will bring rain.[24]

5. The peacock, a lifelong celibate, impregnates the peahen by shedding a tear.[25]
6. The cow is the only animal which 'inhales and exhales oxygen'.[26]
7. The Vedas have a theory that is superior to Albert Einstein's $e=mc^2$ and the theory of relativity.[27]

The right wing also claims that our ancestors were more liberal towards those who took credit for the inventions in ancient India. Such claims are now a growing cultural constituency of right-wing politics in the country and social media is used as a tool to spread this.

It is no more an exaggeration to say that rumour plays a very significant role in contemporary Indian politics and the rise of conservative politics. Spreading rumours becomes more prevalent during political campaigns prior to elections. As the election season approaches, everything is worth a try in the attempt to defeat rivals. Electioneering campaigns are rife with rumours, especially concerning politicians' involvement in scams and honey traps. Rumours are widely used by competitive politicians as part of their efforts to discredit formidable political opponents or political parties.

Article 19 of our Constitution ensures that you have the right to freedom of speech. Spreading unverified and unscientific information on the Internet can get you booked under the Indian Penal Code and relevant sections of the Information Technology Act 2000 (the IT Act 2000). Be cautious so that you are not used as an instrument for spreading fake information just by clicking the forward or share button on your apps. The irony is that, over a long period, plenty of people have simply become preys to gossips and rumours spread on the Internet.

Endnotes

1. 'Fact Check: Report Claims "Rohingyas Are Eating Hindus",' Uses Image of Tibetan Funeral Ritual', *Scroll.in*, 18 December 2018, https://scroll.in/article/906326/fact-check-report-claims-rohingyas-are-eating-hindus-uses-image-of-tibetan-funeral-ritual, last accessed on 24 May 2019.

2. Vidhi Doshi, 'India's Millions of New Internet Users Are Falling for Fake News—Sometimes with Deadly Consequences', *The Washington Post*, 1 October 2017, https://www.washingtonpost.com/world/asia_pacific/indias-millions-of-new-internet-users-are-falling-for-fake-news--sometimes-with-deadly-consequences/2017/10/01/f078eaee-9f7f-11e7-8ed4-a750b67c552b_story.html?utm_term=.cdec115b0f3b, last accessed on 24 May 2019.

3. *The Quint*, https://www.thequint.com/quintlab/lynching-in-india/, last accessed on 15 May 2019.

4. Alison Saldanha, '2017 Deadliest Year For Cow-Related Hate Crime Since 2010, 86% Of Those Killed Muslim', *India Spend*, 8 December 2017, https://www.indiaspend.com/2017-deadliest-year-for-cow-related-hate-crime-since-2010-86-of-those-killed-muslim-12662/, last accessed on 15 May 2019

5. Alison Saldanha, '2017 Deadliest Year For Cow-Related Hate Crime Since 2010, 86% Of Those Killed Muslim', *IndiaSpend*, 8 December 2017, https://www.indiaspend.com/2017-deadliest-year-for-cow-related-hate-crime-since-2010-86-of-those-killed-muslim-12662/, last accessed on 15 May 2019.

6. Dutta, Prabhash K. '16 Lynchings in 2 Months. Is Social Media the New Serial Killer?', 2 July 2018, https://www.indiatoday.in/india/story/16-lynchings-in-2-months-is-social-media-the-new-serial-killer-1275182-2018-07-02, last accessed on 15 may 2019.

7. Shadab Nazmi, Dhruv Nenwani and Gagan Narhe, 'Social Media Rumours in India: Counting the Dead', *BBC*, 13 November 2018, https://www.bbc.co.uk/news/resources/idt-e5043092-f7f0-42e9-9848-5274ac896e6d, last accessed on 15 may 2019.

8. Alison Saldanha, Pranav Rajput and Jay Hazare, 'Child-Lifting Rumours: 33 Killed In 69 Mob Attacks Since Jan 2017. Before That Only 1 Attack In 2012', *IndiaSpend*, 9 July 2018, https://www.indiaspend.com/child-

lifting-rumours-33-killed-in-69-mob-attacks-since-jan-2017-before-that-only-1-attack-in-2012-2012/, last accessed on 15 May 2019.

9. IndiaSpend, 'Child-lifting Rumours Caused 69 Mob Attacks, 33 Deaths in Last 18 Months', 9 July 2018, https://www.business-standard.com/article/current-affairs/69-mob-attacks-on-child-lifting-rumours-since-jan-17-only-one-before-that-118070900081_1.html, last accessed on 15 May 2019.

10. Alison Saldanha, '2017 Deadliest Year For Cow-Related Hate Crime Since 2010, 86% Of Those Killed Muslim', *IndiaSpend*, 8 December 2017, https://www.indiaspend.com/2017-deadliest-year-for-cow-related-hate-crime-since-2010-86-of-those-killed-muslim-12662/, last accessed on 15 May 2019.

11. P Anima, 'Call to Kill', *Business Line*, 20 July 2018, https://www.thehindubusinessline.com/blink/cover/call-to-kill/article24469984.ece, last accessed on 15 May 2019.

12. https://docs.google.com/spreadsheets/d/1eHrxOdd08BoDYT1E2hKOTNyhNBSqrj0u5p7s2CDZKwU/edit#gid=0

13. Sam Wineburg, Sarah McGrew, et. al., *Evaluating Information: The Cornerstone of Civic Online Reasoning. Stanford Digital Repository*. https://purl.stanford.edu/fv751yt5934, 2016, last accessed on 15 May 2019.

14. Camila Domonoske. 'Students Have "Dismaying" Inability To Tell Fake News From Real, Study Finds', *NRP*, 23 November 2016, https://www.npr.org/sections/thetwo-way/2016/11/23/503129818/study-finds-students-have-dismaying-inability-to-tell-fake-news-from-real, last accessed on 17 May 2019.

15. Storyful Intelligence, 'India and Misinformation', https://storyful.com/thought-leadership/india-and-misinformation/, 1 October 2018, accessed on 17 May 2019.

16. Meera Nanda, 'A Detailed Account of This Issue Is Documented', *Science in Saffron: Skeptical Essays on History of Science*, New Delhi: Three Essays Collective, 2016.

17. Christina Zaho, 'Minister Ridiculed for Saying Internet Was Invented by Ancient India Thousands of Years Ago', *Newsweek*, 18 April 2018, http://www.newsweek.com/minister-ridiculed-saying-internet-was-invented-ancient-india-5000-years-ago-890567, last accessed on 20 April 2018.

18. Maseeh Rahman, 'Indian Prime Minister Claims Genetic Science Existed in Ancient Times', *The Guardian*, 28 October 2014, https://www.theguardian.com/world/2014/oct/28/indian-prime-minister-genetic-science-existed-ancient-times, last accessed 20 April 2018.
19. In the Indian Science Congress held January 2015 in Mumbai, Capt. Anand J. Bodas, the retired principal of a pilot training facility, claimed the world's first plane was invented by the Hindu sage Maharishi Bharadwaj, see, Rama Lakshmi, 'Indians Invented Planes 7,000 Years Ago—And Other Startling Claims At the Science Congress', *The Washington Post*, 4 January 2015, https://www.washingtonpost.com/news/worldviews/wp/2015/01/04/indians-invented-planes-7000-years-ago-and-other-startling-claims-at-the-science-congress/?utm_term=.48a44ed7130d, last accessed on 20 April 2018.
20. Reuters, 'Ancient India had Aeroplanes, Nuclear Weapons, Says Chief of India's Premier History Body', *India Today*, 21 November 2014, https://www.indiatoday.in/india/story/ancient-india-aeroplanes-nuclear-weapons-chief-history-chairman-indian-council-historical-research-228149-2014-11-21, last accessed on 20 April 2018.
21. Yogi Agaarwal, 'RSS Agenda: Replacing Science With Myths', *Deccan Chronicle*, 30 January 2018, https://www.deccanchronicle.com/opinion/op-ed/300118/rss-agenda-replacing-science-with-myths.html, last accessed on 20 April 2018.
22. Michael Safi, 'Indian Education Minister Dismisses Theory of Evolution', *The Guardian*, 23 January 2018, https://www.theguardian.com/world/2018/jan/23/indian-education-minister-dismisses-theory-of-evolution-satyapal-singh, last accessed on 20 April 2018.
23. Participating in a discussion on the Union budget, the Appropriation (Vote on Account) Bill, 2015 and the Appropriation Bill, 2015, ShankarbhaiVegad, a BJP MP from Gujarat in the Rajya Sabha said Dung and urine from cows are more potent than medicines and can cure deadly diseases like cancer, see, Smriti Kak Ramachandran, 'Cow Dung, Urine Can Cure Cancer: BJP MP', *The Hindu*, 19 March 2015, http://www.thehindu.com/news/national/cow-urine-can-cure-cancer-bjp-member-in-rs/article7012010.ece, last accessed on 20 April 2018.
24. In 2016, Viredner Singh, BJP MP from Badhoi in Uttar Pradesh, Virendra Singh remarked in the Lok Sabha that ritual burning of ghee and other food stuff would bring rains, see, PTI 'Yagya Brings

Rain, Produces 300% More Oxygen: BJP MP in Lok Sabha' *Hindustan Times*, 5 August 2016, https://www.hindustantimes.com/india-news/yagya-brings-rain-produces-300-more-oxygen-bjp-mp-in-lok-sabha/story-RwC6ON57COkRFi6MUgchyO.html, last accessed on 20 April 2018.

25. On June 2017, former Justice Mahesh Chandra Sharma of the Rajasthan High Court suggested in an interview that India's national bird, the peacock, 'is a lifelong celibate' that impregnates the peahen by shedding a tear. He cited Lord Krishna's use of a peacock feather as proof of its celibacy, see, FE Online, 'Peacock Is a Lifelong Celibate': Tweeple Go Crazy over Rajasthan HC Judge's Remark; Here Are Some Funniest Rweets, *Financial Express*, 1 June 2017, https://www.financialexpress.com/india-news/peacock-is-a-lifelong-celibate-tweeple-go-crazy-over-rajasthan-hc-judges-remark-here-are-some-funniest-tweets/696202/, accessed on 20 April 2018.

26. Rajasthan education and Panchayati Raj Minister Vasudev Devnani has said that cow is the only animal that 'inhales and exhales oxygen' and that people need to understand its 'scientific significance', see, HT Correspondent, 'Cow Only Animal To Inhale and Exhale Oxygen: Rajasthan Minister', *Hindustan Times*, 16 January 2017, https://www.hindustantimes.com/india-news/cow-only-animal-to-inhale-and-exhale-oxygen-rajasthan-minister/story-a8nPi8XDxpvO8YKwibN5RJ.html, last accessed on 20 April 2018.

27. Science and Technology Minister Harsh Vardhan on March 2018 claimed that cosmologist Stephen Hawking, had said the Vedas have a theory that is superior to Albert Einstein's $e=mc^2$ and theory of relativity, see, PTI, 'Stephen Hawking Said the Vedas Have Theory Superior To Albert Einstein's $e=mc^2$: Harsh Vardhan', *The Indian Express*, 16 March 2018, http://indianexpress.com/article/india/stephen-hawking-said-vedas-have-theory-superior-to-albert-einsteins-emc2-harsh-vardhan-5100099/, accessed on 20 April 2018.

4

Hate Campaign

Mehdi Masoor Biswas, a Bengaluru-based twenty-eight-year-old engineer would seem an ordinary person at the outset. He shares jokes, images and pictures of food, especially pizza, on social media platforms. Otherwise shy and introvert, his Twitter handle @ShamiWitness landed him in prison.[1] In just a couple of years, he had posted over one lakh tweets, an average of a hundred tweets a day, sharing it with over seventeen thousand followers. He actively worked for the cause of the Islamic State (IS)—a multi-country terror outfit—retweeted, shared and even translated Arabic posts that support the IS phenomenon, using an online translator. This is a typical case of double life made possible—one online and the other offline.

He was a loner and a recluse, but like many introverts, social media was his place of comfort. On social media, he was actively supporting the cause of the Islamic State, defended beheadings, motivated youths to fight for it, 'made fun of the news of rape of Kurdish women'.[2] However, those who knew him were shocked to see this side of him.

He was accused of supporting Islamic terrorism and

sympathizing with the cause of the Islamic State through his activity online—a crime still not properly defined in Indian cyber laws. Under Section 66f of Information Technology Act, what he had been doing was insufficient to be defined as cyber terrorism. While he was promoting IS activities or beliefs in virtual space, he was also a law-abiding citizen offline. The police had been looking for evidence to establish a link between his online activities and his behaviour offline. His activities were neither simple nor easily identifiable for the policing system. His arrest raged an intense legal battle in India. The battle included the question whether sympathizing with a 'cause' on social media should be labelled as an act of terror in the Indian Court of Law. It was for the first time that the legal and police establishments in India encountered the problem of the thin line dividing the offline and online life of a person, and whether there is any connection between instigating people online and criminal activity offline. The connection with the real world cannot be established yet, other than a Twitter account to prove his involvement in a terror case. He openly admitted his support for the ideology of the Islamic State and their cause, but he had no connection in the real world with them. He was not an enrolled or registered member of the Islamic State nor was there any evidence to prove that he was involved in any other real activity with them. Indeed, this was a case study, which stormed into a debate around hate speech on social media. Whether free speech should be limited on social media caught the attention of policymakers. What is the extent of sympathizing with a cause and when does one's involvement turn into a crime required to be defined?

Social media has acted as a catalyst to instigate revolutions, topple political regimes and bring about radical social changes across the world. At the same time, militants, communalists, casteists, racists, terrorists, misogynists, fanatics, gangsters and extremists increasingly use these amazingly potent techniques. If left unregulated, these can facilitate divisive groups to spread violence, hatred and crime, while it brings in more crises.

Hate groups robustly use the Internet to spread hatred and recruit new members. Social media is a powerful medium to communicate radical ideas to a larger audience. It is easy to spew venom on anyone as long as there is an unregulated space on the social media platform.

The Origins of Hate Groups on the Internet

The first hate group took to the Internet in 1995. Today, an astounding array of extremists finds expression on the Internet. Extremist groups were one of the early adopters of electronic communication tools to spread their message in a more effective way. For example, in 1985, long before most people had heard about the Internet, Tom Metzger, the leader of the White Aryan Resistance, created a computer bulletin board,[3] and the group has had an active online presence ever since.[4] White supremacists were one of the early adopters of electronic communication technology to spread their messages. For example, in 1995, Don Black—a former leader of an American white supremacist hate group Ku Klux Klan, commonly called the KKK or the Klan—created Stormfront, the first major white supremacist website.[5] Today, many other racist, religious, ideological and other extremist sites, similar to Stormfront, exist on the Internet. However, the rise of social media has opened more avenues for hate groups to establish their presence.[6] There has been a rapid increase of online hate groups, both from the Left and Right, from white supremacists to eco-terrorists and trans-national jihadists in the European context.[7]

There is a variety of writings on hate speech, and it is

defined in a variety of ways. In the literary sense, any abusive or threatening speech or writing against a particular person or group, especially race, colour, ethnicity, gender, disability, sexual orientation, nationality or religion can be termed as hate speech. Articulated in forms of speech, gesture or conduct, writing or illustrations, it usually shows incitement to violence or spurring prejudice against an individual or group. A hate speech online on the Internet platforms is even more dangerous.[8]

Hate speech on the Internet is often understood to involve the intentional infliction of substantial emotional distress to someone or any group. It tries to hurt people. What it tries to belittle is the faith, culture, religion, caste, gender, sexuality and even the language and class of people. The Law Commission of India reports that hate speech amounts to instilling fear or alarm, or inciting violence.[9] It may involve verbal, image, audio, video and text-based abuses or attacks targeting a community. Hate speech can also contain content that may not necessarily be abusive in nature, but is sufficient to incite violence against a particular section of the society. It is a deliberate attempt to attack the reputation of an individual or group of people and bring shame to them. In the end, hate speeches cause mental stress and tension in the minds of people.

At the same time, little is known about it from the prism of hate speech laws such as the Information Technology Act. The Constitution of India is enshrined with the prevention of hate speeches. Several sections of the Indian Penal Code and the Code of Criminal Procedure enumerate the same. The relevant sections of the Information Technology Act are also used against hate speeches on the Internet.

In the Indian context, hate campaigns on the Internet have a different dimension, since it can be connected to a number of social variables such as gender, religion, caste, faith, language, region, sexuality and culture. So, any attempt to bring hate under control must take into consideration the cultural diversity in the country. Since the cultural particularities of hate speech differ from Western and non-Western cultures, the nature and content of hate also differs. There are different types of hate speech on the Internet, which caters to the social divides in India. Moreover, as the Internet provides ample opportunities to publish content even from fake profiles, it makes the issue a little more complicated.

Cybercrimes

Another type of hate speech is based upon faith. There are a number of groups, mostly unheard-of, who start defending some faith as superior. Muslims are portrayed as anti-nationals and Christians as against the Indian culture. However, Vedic knowledge is considered as scientific. In the eyes of such fanatics, those who had migrated to other religions in the past were tagged as enemies of Bharat. To justify their claims, they distort history, misinterpret facts and claim Vedic knowledge as supreme. They spread hate gradually. Such hate speeches spread among those who are unable to distinguish fact from rumour; they fall prey to the emotionally charged appeals on cyberspace.

The Internet is now heavily capitalized by Neo-Brahmanic forces. It has enormous intellectual and 'cultural capital' base in cyberspace, which attempts to recuperate the Brahmanic supremacy. In the far end of the 1990s, when I was an undergraduate student, I still remember friends looking at people with contempt, who believed in the ideologies of the Sangh Parivar. They were kept at a distance because people believing in progressive ideas thought that Sangh's ideas are retrogressive. That said, in those days and even before, people never openly admitted supporting the Sangh ideology. Still, I remember friends spreading rumours against teachers supporting right-wing ideas. Now, people have started openly supporting Sangh activities in cyberspace. Information related to right-wing ideas are shared in plenty. This is certainly a cultural capital for neo-Brahmanic forces.

In the past, the ruling establishments patronized it. Since the ideology cannot be reinstated in a democratic liberal social order, which has the backing of a rational constitution, divisive ideas regain their ground only by using culture as a weapon of legitimacy. It calls for cultural and intellectual slavery from the 'other', who belong to the Brahmanic social order. Now, the upper-caste Hindu Brahmanic superiority is trying to create their legitimacy in cyberspace. It is made possible by strategic use of online propaganda. They say that temples are targeted by those who hate Hindus; Hindu women are being targeted by Islamists and are lured by them into 'love traps'; Bollywood stars are all Muslim men, and their girlfriends and wives are Hindus; all terrorists are Muslims and they target India; and Muslims are all set to become a majority over Hindus. These claims in cyberspace find a huge audience. However, those who fall for such hate-filled campaigns are unknowingly providing their services to the Brahmanic order to legitimize their hold over the people. The wider acceptance of this sort of idea is found among users who are unable to spot the bias of extremist groups. Thus, the cultural power of Brahmanic forces is ignited. The Internet is the site of this legitimacy-milling factory. Neo-Brahmanic forces use it as a tool for propaganda to serve their interest best by indoctrinating the underclass of the caste hierarchy. Dogmas are now being legitimized.

Ordinary users of the Internet, particularly those on Facebook, Twitter, YouTube and other social media sites, who are easily being influenced by rumours and gossips on the

Internet, are unknowingly providing their services to neo-Brahmanic forces. For example, there are people who share sexist content that substantiates the said idea, for instance, women should be barred from entering temples during their menstrual cycles. There is no reason as to why women should not enter temples during their menstrual cycles. People devoid of such reasoning are vulnerable and easily fall prey to the tact of the divisive creators. They also become a tool in the hands of people who use propaganda to support divisive ideologies and belief systems. They don't understand that women are humans, should be treated with respect and have the right to be treated equally. What people do see on the Internet is only one type of content—a patriarchal, feudal social system and cultural capital being reinstated, and they are gaining legitimacy by way of constant online propaganda. In addition, people unknowingly endorse this kind of information.

The ordinary people often hail political leaders and broadcast their speeches by tagging them as 'Yugapurush'—man of the age. Such lofty comments and the large number of page views on their updates are likely to have other vested interests for appearing in cyberspace. The same is for celebrities and youth icons as well. Indeed, the ordinary people are fooled by such strategic use of social media. For example, some people share content against minorities and those on the periphery of the society in cyberspace. In doing so, they show their casteist mindset and patriarchal attitude. People find news about inter-caste affairs or sexual minorities more interesting and such news become viral and start trending on social media. However, this sort of interest being generated on the Internet caters to a particular kind of ideology—the right wing, feudal social system in which god, men, the Vedas, and superiority of the so-called 'ancient India' are of prime importance.

People get easily attracted to online content that glorify the sanctity of the Vedas and uphold the authenticity of ancient science. They support the idea of religions and faith above the rule of law and the Constitution of India. At the centre of this remains that old assumption—Brahmanism and its contrived superiority in the Indian culture. However, the fact is that those who serve as the mouthpieces of the Brahmanic supremacy are not its benefactors. In fact, they are its victims. The irony is that the prey are influenced into intellectual slavery by their predators on the Internet. The content they share and prefer in cyberspace is not invented by those who share it. A group of people with wicked intention invent them. Moreover, a majority of such likes and shares are not a part of the twice-born, rather, they belong to the lower caste. The paradox is that cyberspace is used as a site to build legitimacy. Hate is deliberately manufactured in cyberspace.

The Internet is enabling a community of men who want to ruin women's lives. Hate against women is rampant on the Internet. Often, men, who are absolute strangers, follow women on social media, send friend requests, obscene messages and photos, unnecessarily like and share their posts. When the women do not reciprocate, the men abuse them. They are labelled as ugly, whores, nymphomaniacs, sluts and characterless. Abuses multiply on their social media pages. They are stalked, harassed and threatened. Fear is instilled by way of threats of rape and acid attacks. This is in stark violation of many laws.

A noted transgender activist in south India put a note on her Facebook that a Sri Lankan citizen has been stalking her, since she has declined his proposal. He has begun to abuse her on Facebook and other social networking sites where she has an account. The point is that transgender women are easy bait to cyberstalking. This is not an isolated incident. Lesbians, gays, intersex, bisexuals, etc., also face this problem. In the online spaces, such people misuse the freedom of expression to abuse and spread prejudices against other people based on their sexual orientation. Hate against sexual minorities is trending on social media. Many online platforms, especially chat rooms, are replete with prejudices and stereotypes against the sexual minorities. Sexual minorities are often bullyied, harassed and threatened on the Internet. The LGBT community has to confront the widespread bias against them and that they are portrayed as licentious, hedonistic and AIDS-spreading aliens. Tamil playback singer Chinmayi Sripada was trolled due to her stand on the Tamil Fishermen issue among other things. On 18 October 2012, Chinmayi Sripada (@Chinmayi), the singer and one of the most conversational celebrities on Twitter, lodged a complaint stating that a few people have been tweeting about her and her mother and making 'casteist' and 'vulgar' remarks. The comments were posted by Twitter handles @rajanleaks, @enthilchn, @sharavkay, @losongelesram, @vivajilal and @thyirvadai. According to her statement to news sources, the problem started two years ago after they included her in a hashtag on a fishermen's issue and wanted her to support the cause. She did not entertain the request, as there were vulgar comments about many top leaders.[10] Chinmayi tried the usual way of blocking the offenders initially. It didn't end there. New people jumped into the scene and she was known soon as the famous singer who blocked anyone who tweeted in Tamil. This turned into a campaign because it was assumed that Chinmayi was against Tamilians on Twitter.

Ramachandra Guha, the historian, has reportedly faced trolls by the right-wing forces for his comments regarding history. Meena Kandaswamy, a prominent Dalit feminist poet and writer, was subjected to a similar stream of abuse on Twitter, following her support for the beef festival being organized on the premises of Osmania University in Hyderabad. The Internet has enormous examples of people facing hate attacks instigated by perceived notions of cultural superiority. They claim their language, music, dance forms and other cultural artefacts as superior. They try to disturb people, who uphold different views from the mainstream and official versions.

The Internet is furthering caste divides. Caste-based discrimination is found in plenty on the Internet. Indeed, the Internet has reproduced a Hindu social order. There are countless incidents where people have been threatened of grave consequences by often anonymous and conservative trolls who post outrageous libel to evoke emotional responses about caste. A conservative faction, a sort of a surveillance mechanism, is systematically organized for such operations. This is just the tip of the iceberg. Hate on the Internet and its implications are manifold in India.

Anatomy of Online Hate

The abusive behaviour of certain people in cyberspace has diluted the Internet as a free space for all, devoid of any kind of divisions. Since the personality of the Internet is such that people are free to air their opinions—right or wrong—freedom to publish has, in fact, maligned the Internet and refurnished cyberspace as a haters' paradise.

People who seem secular offline become communal in cyberspace. They remain secular in the day, and by the night, they become communal. Educated people who are employed in multinational firms with good salary and are well connected, all of a sudden, are drawn towards extremism. Some people use fake accounts to spread communal ideas and network with extremist groups on social networking sites. Some people, who are known to be liberals among friends, turn into religious fanatics in cyberspace.

Even introverts feel empowered by cyberspace and what they spread is full of hate. For vindictive people, the Internet is a propaganda tool. It is the cultural capital for the haters. The Internet has helped many people enforce their communal mindset, which is, otherwise, hidden under the pretext of secular institutions and social vigilantism.

Diffusion of hatred on cyberspace is not a random incident carried over suddenly by some wicked minds. It is a planned execution of an agenda, the nature of which is yet to be identified. However, its direction is clear that it tries to establish a polity based on the purity of ancient India. Hate doesn't appear so quickly. Like any communal riot, which breaks out after a long gestation period, hate in cyberspace has its own alchemy unique to the personality of the Internet.

Hate Campaign on Facebook

Although Freethinkers[11] claim to promote 'science and rationalism', a majority of its members are drawn from the right-wing conservative forces, which raucously post prejudices in the online space. Its YouTube page has over two thousand subscribers and Facebook has over sixteen thousand members. There are discussions, for example, about the Taj Mahal, with many claiming it to have been built on a destroyed Hindu temple. Some discussions deny that the Taj Mahal is a symbol of love; some say that the Mughal Emperor Shah Jahan took Mumtaz Mahal as his fifth wife after killing more than just a few people.

The members of hate groups promote ideas in stark opposition to Article 51A(h) of The Constitution of India, which states that it is the duty of citizens to develop a scientific temper, humanism and the spirit of inquiry and reform. Unverified information perpetrated as facts gather like endless storms. Young people exposed to these false pieces of information easily believe them without even checking their authenticity. More importantly, discussions of such a nature are taking place on the pages of groups on Facebook that says its objective was to promote science and rationalism.

How to Identify Hate Campaigns

There are websites, social networking groups, web portals, blogs, chat rooms, videos and hate games that promote racial violence, anti-Semitism, homophobia, hate music and terrorism. The world over, it is believed that tens of thousands of online hate platforms operate.[12]

A study by the Internet Security system, Websense, claims higher levels of dispersion of 'hate and militancy' sites in recent times. It also reports that racism, hate and radicalism in the online space have tripled.[13] In particular, they have reported a substantial increase of the same on social networking sites and platforms such as YouTube, Yahoo!, Google+, etc. A hate speech is not easily traceable on the Internet until it has been reported. There are instances of screening agencies logging on to hate websites and removing objectionable content. Then, the hate groups quit the site and move on to other sites, using different service providers, often from another country. There are some methods used for spreading hate on the Internet. A study[14] by the British Institute of Human Rights (BIHR) for the Council of Europe classifies hate speeches on the basis of hate sites, blogs and online forums, emails and personal messages, gaming, social networking sites, videos, music and automated content, astroturfing and fictitious identities, and shows where they are most frequent on the Internet.[15]

For instance, in 2013, the Government of India blocked over 250 websites and social networking sites accused of spreading inflammatory content that incited panic among thousands of workers and students from the country's north-eastern states.[16] Roughly, 20 per cent of the webpages blocked by the government agencies were administered by conservative establishments which were trying to polarize the country on communal ranks. The list also consisted of URLs of mainstream media websites.[17]

Text messaging through email and apps, such as WhatsApp, is another source of hate speeches. Private emails or personal messages are the hardest to control. Emails used to back supporters and broaden ideology, in a private space, are almost impossible to monitor. Personal messages are also used to target and intimidate individuals. The texting technology allows people to send messages anonymously to people.

A good example to cite here would be the scandal, in which a few top Australian police officers were embroiled. Some police officers of the Australian state of Victoria had circulated what the press there called a 'sickening video footage', showing the death of a man who was travelling on the roof of a crowded train in India. The video was accompanied by a comment that said electrocution could be 'a way to fix the Indian student problem' in Melbourne. The e-mail, containing the video, was circulated within the Victoria Police computer system, and many racist comments were added to it.[18] With the advent of social networking sites, such as Facebook, racist and bigoted opinions and recruitments by trollers have increased in recent years by hate groups.[19] The Social Networking Sites (SNSs) community can contrive insidious propaganda.

Hate music and videos are frequently used to draw supporters and often to raise funds for racist groups. YouTube, Vimeo,

MetCafe and MySpace are increasingly being used. For example, Akbaruddin Owaisi, an MLA of All India Majlis-e-Ittehadul Muslimeen (MIM), was booked *suo motu* for allegedly using inflammatory and derogatory language against a community during his public speeches by the Nizamabad II Town police and the Nirmal police in Adilabad and Nizamabad, besides the Osmania University police in Hyderabad and other parts of Andhra Pradesh.[20] In just a few days, the number of views his videos had garnered gives a sense of the fast rising spread of violence via video sites. On one hand, online harassment and hate campaigns are turning hydra-headed and harming the real democratic spirit of the Internet. Little attention has been furnished by the political leadership and the civil society regarding the vulnerability of ordinary people falling prey to such campaigns. And, in spite of the frequency of such campaigns, those behind the crime remain invisible, operating as anonymous and fake profiles. They will reappear with new profiles once they get blocked. As the Internet provides enormous possibilities of sharing information, it is often used for spreading hate and diffusing violence in the society. On 2 October 2013, I had organized an ethnographic observation on the online platforms of *Hindustan Times*, *The Pioneer*, *Deccan Herald* and many other mainstream media, and I noticed that there was no comment moderation policy on the websites. Even if there were moderation and screening policies, one can easily get their replies and comments published by way of fake identities and email addresses.

At the National Integration Council on 23 September 2013, Akhilesh Yadav, the then chief minister of Uttar Pradesh (UP) hinted at the fact that social media is used by miscreants in order to vandalize society and spread hate in India. He was referring to the communal clashes of 2013 in Muzaffarnagar, UP. Communal riots broke in Muzaffarnagar on 7 September 2013 and more than fifty people had died leaving over 40,000 homeless.[21]

Hate groups have been using social networking sites to buff communal tensions and beleaguer people from the northeastern Indian states residing in cities like Hyderabad, Bangalore, Pune and Mumbai.

Similar apprehensions, about social media platforms being used to spread hate speeches, were raised in the Parliament in August 2012. There were stories of morphed pictures being used in the form of multi-media messages (MMS) with a vicious motive to incite tension and hate.

The Government of India decided to ban selected social networking sites and the sending of bulk MMS and SMS for fifteen days, across the country from 17 August 2012, anticipating hate campaigns on the Internet at that time. The unpleasant incidents resulted in more than 1,500 people from northeast India, especially students, leaving Maharashtra in just four days.[22]

The government attempted to remove the objectionable content from the Internet sites by requesting the companies who operate them. However, it is difficult to do so as these morphed pictures and videos are uploaded on the websites from different countries, often from those that do not have cordial diplomatic relations with India. Simultaneously, the US-based social media websites interpret privacy and freedom of speech in a different manner, which is in contrast to the laws existing in India.

Therefore, when objectionable content on the Internet becomes a diplomatic issue for the government, it becomes more difficult for the government to deal with the users.

As technology grows, freedom gets a comparatively newer dimension in which excesses and extremes are likely to occur. Such a freedom is always at a risk of being abused. The freedom may be compared to the sudden release of a person kept under strict regulations for a long time—he may act intemperately. This is true of some specific and sedate patterns of the use of social media in India. Apparently, hate speeches on the Internet has its ancestry in the bias, bigotry and prejudice of our society, and this existed long before the advent of the Internet.

The presence of online hate groups, in an environment frequented by the youth, is a potentially dangerous combination. Research shows that various online communities and social networking sites offer sources of social identification for the youth and many young people do not distinguish between the people they meet online from those offline.[23] Thus, hate groups, even those engaging in the virtual world, can become important socializing agents in the lives of these youth. Exposure to online hate ideology thus occurs relatively frequently. While it is undeniable that hate groups existing on the Internet are highly active, they also actively target the youth and their messages reach a significant number of young people. A perusal of the Internet reveals that the chances of young people engaging in hate-inspired activities are likely to increase in the days to come. Social media is like a Speaker's Corner in Hyde Park, London, which is famous for allowing free speech. Here, people are permitted to speak and debate, share ideas and hold discussions. There is nothing clearly proscribed and all the people are generally free to speak about anything to a growing audience. In a multi-ethnic society like India, with millions of people from different religions, castes, dialects and regions, social media is a very sensitive area. The motive of the use of this example here is to raise an analogy and draw your attention to a very simple question: Is the social media analogous to the Hyde Park of London providing unheard voices opportunities to be heard? The interesting thing about the Internet is that the speakers at the park are veiled by the obscurity of a virtual identity. The speech turning into an unfortunate incident, or turning hostile, and social conflicts due to this culture of open speech is more likely, which certainly dampen democracy.

Endnotes

1. Devesh K. Pandey, Aditya Bharadwaja and Shiv Sahay Singh, 'The Case Of a Virtual Warrior', 28 December 2014, https://www.thehindu.com/sunday-anchor/the-case-of-virtual-warrior/article6731205.ece, accessed on 15 May 2019.
2. Ibid.
3. M. Hamm, *American Skinheads: The Criminology and Control of Hate Crime*, Westport, CT: Praeger, 1993.
4. B. Levin, 'Cyberhate: A Legal and Historical Analysis of Extremists' Use Of Computer Networks in America', *American Behavioral Scientist*, 45 (6): 958–98, 2002. DOI: 10.1177/0002764202045006 03.
5. High Tech Hate: Extremist Use of the Internet, Anti-Defamation League, Nexis, Stormfront, 1997, https://www.stormfront.org/forum/, last accessed on 15 may 2019.
6. T. Kiilakoski and A. Oksanen, 'Soundtrack of the School Shootings Cultural Script, Music and Male Rage', *Young: Nordic Journal of Youth Research*, 19(3), 2011, pp. 247–69.
7. C. Brown, 'www.hate.com: White Supremacist Discourse on the Internet and the Construction of Whiteness Ideology', *The Howard Journal of Communications*, 20(2), 2009, pp. 189–208 and H. Chen, W. Chung, J. Qin, E. Reid, M. Sageman, and G. Weimann, 'Uncovering the dark Web: A Case Study of Jihad on the Web', *Journal of the American Society for Information Science and Technology*, 59(8), 2008, pp. 1347–59.
8. See works, Reiter Rachel Weintraub, 'Hate Speech over the Internet: A Traditional Constitutional Analysis or a New Cyber Constitution?', 8 B.U. PUB. INT. L.J. 145, 149 (citing Samuel Walker, Hate Speech: The History of an American Controversy 8 (1994), 1998; Richard Delegado and Jean Stefancic, 'Must We Defend Nazis? Hate Speech, Pornography, and the New First Amendment', *New York: New York University Press*, 1997; Michel Rosenfeld, 'Hate Speech in Constitutional Jurisprudence: A Comparative Analysis', *Cardozo Law Review*, 1523, 1523, 2003.
9. https://www.firstpost.com/india/hate-speech-in-india-medias-rabble-

rousing-doesnt-help-cause-proves-counter-productive-to-free-speech-5182231.html
10. Desh Kapoor, 'Police Arrest Twitter Users for Harassment of a Singer in Chennai', *Patheos*, 23 October 2012, https://www.patheos.com/blogs/drishtikone/2012/10/police-arrest-twitter-users-for-harassment-of-a-singer-in-chennai/, accessed on 15 May 2019.
11. facebook.com/groups/ftkerala6/
12. 'Facebook Hate Groups Target Jews and Gays', *Pink News*, 15 May 2009, web, http://www.pinknews.co.uk, accessed on 07-02-2013.
13. 'Racism, Hate, Militancy Sites Proliferating via Social Networking', *Networkworld*, May 2009, http://www.networkworld.com/news/2009/052909-hate-sites.html, accessed 02 January 2013.
14. 'Mapping Study on Projects Against Hate Speech Online', Council of Europe, *British Institute of Human Rights*, 2012.
15. Clay Calvert and Robert D. Richards, 'New Millennium, Same Old Speech: Technology Changes, but the First Amendment Issues Don't', 79 B.U. L. REV, 1999, pp. 959–60.
16. Rama Lakshmi, 'India Blocks More Than 250 Websites for Inciting Hate, Panic', *Washington Post*, web, 20 August 2012, www.washingtonpost.com, accessed on 4 February 2013.
17. PraneshPrakash, 'Analysing Latest List of Blocked Sites' (Communalism & Rioting Edition), *The Centre for Internet and Society*, 22 Aug 2012, http://cis-india.org/internet-governance/blog/analysing-blocked-sites-riots-communalism, accessed 28 December 2012.'
18. 'India Outraged at Aussie Cops' "sick hate email"', *The Indian Express*, 10 October 2010, http://www.indianexpress.com, accessed on 04 February 2013.
19. 'Hate Groups Increasingly Use Social Networking to Recruit', *Fox News*, 14 May 2009, http://www.foxnews.com, accessed 07 February 2013.
20 PTI, 'Akbaruddin Owaisi Booked in One More "Hate Speech" Case', *Hindustan Times*, 5 February 2013, https://www.hindustantimes.com/india/akbaruddin-owaisi-booked-in-one-more-hate-speech-case/story-6YD20jj0eLKnRRRMdoiHHP.html, last accessed on 15 February 2013.
21. India Today Online, 'Muzaffarnagar Violence: After Rajnath, Congress Also Blames SP for Riots in Uttar Pradesh', *India Today*, 21 September 2013, https://www.indiatoday.in/india/north/story/

muzaffarnagar-violence-after-rajnath-congress-also-blames-sp-for-riots-in-up-211803-2013-09-21, last accessed on 15 May 2019.
22. '1500 North East People Fled Maharashtra in 4 Days', *Bihar*, 17 Aug 2012, *Prabha*, http://news.biharprabha.com/2012/08/1500-north-east-people-fled-maharashtra-in-4-days/, accessed 29 December 2012.
23. V. Lehdonvirta and P. Räsänen, 2011, 'How Do Young People Identify With Online and Online Peer Groups? A Comparison Between UK, Spain and Japan', *Journal of Youth Studies*, 14(1), 2011, pp. 91–108.

5

Internet Trolling

Trolling has transformed the Internet. It is an act of highlighting or mocking a faux pas or an error, to inform the people, educate someone else or reform someone's thinking. A benchmark incident to consider in this regard was the usage of the phrase 'cattle class' by the diplomat-turned-politician Shashi Tharoor in 2009 on Twitter. Ever since trolling acquired huge public acceptance as a means of social vigilantism, it has brought about some positive impacts. It has made politicians conscious about their public conduct. It has made stars, youth icons and public figures responsible to some extent.

Internet trolling now tends to show the narcissistic mindset of the society. Trolling is ruining the Internet landscape. In July 2018, twenty-one-year-old Hanan Hamid became a victim of intense trolling on social media for selling fish in Ernakulam, Kerala, after college hours to fund her studies and make ends meet. She was a B.Sc. student at a private college. Her story was carried by leading Malayalam daily *Mathrubhumi*. Many people, including film artists and politicians, shared her touching story on their profiles. However, a section of social media called

Hanan's story a fake. The student faced vicious cyber attack. In between the trolling incidents of Hanan and Tharoor, the Internet shows a wide range of social transformation in our country. All of a sudden, some unknown people pop up, act 'macho' to infuriate others, scan whatever others have posted and question their social media profiles. They express their difference of opinion about something posted on the Internet, or they label a deed or someone's statement as unacceptable. The use of some words, comments and opinions upset them, which was not meant to harm anyone. They get into heated arguments; even take extreme steps to prevent similar things from happening, which they perceive as objectionable, from repeating. They are not even bothered about the trauma created by such unwanted attack, nor are they willing to engage in any kind of decent dialogue. In doing so, such people assume that they have all the right to interfere in the privacy of a person and question them on social media. Even people who have remained anonymous so far intervene in things they think they have all the right to and become the self-appointed authorities of what is being discussed. This means they consider themselves to have all the right to protect it. They function as a well-trained cyber army, who believe in certain ideologies, unquestionable leaderships and authority of some belief systems. Their strange behaviour points to the fact that the rise of social media has led to a strange new discovery. It is the strong presence of Internet trollers. The term is used to refer to user profiles on the Internet who use social media accounts to teach someone a lesson, character assassination and harm someone's reputation. They are the fans of some ideologies, the nature of which is often a point of contest. They consider many things unquestionable, which people are not allowed to have opinions about. They disturb people by a series of cyberattacks. Every morning, they scan the Internet for tracking those who speak against their leaders perceived as undisputed, the Vedas, Manusmriti, communalism, acharyas, among other things. There are things that the Internet trollers and other groups, who are said to be propagators of some sort of belief systems, find obnoxious. By the evening, they share trolls against the so-called liberals and freedom lovers.

Let us first understand who a troller is. A troller on the Internet is someone who starts arguments or upsets people by extraneous messaging with the intent of provoking readers to respond. Their goal is to prey on people's emotions and incite fear in them.

Trolling is anything that claws into broadly two things: the liberal space in society and the secular minds of the people, disturbing people who live peacefully in their world. Most desperate and dissatisfied people would denigrate their perceived enemies on cyberspace for having an opinion or a skill. Lesser confident, illiterate, criminal-minded people annoy others who have a good sense of humour, intelligence and love for fellow beings. Their main goal in life seems to cause confusion, irritate people and make inflammatory statements to get attention.

The etymology of the term is a matter of dispute. Some say that it comes from the old French troller, which means to wander or hunt for a game with no specific target or purpose. It is also used in fishing communities, where trolling is the act of trailing a baited line behind a moving boat, waiting for a fish to catch it. Others believe that the term, troll refers to the Scandinavian mythological beast that lives under bridges, accosting hapless passers-by. The myth says that the troll or the beast that lives under the bridge demands a fee or an answer to its question from the passer-by who wishes to cross the bridge. If the passer-by is able to answer its question, he/she is allowed to cross the bridge.

A troller on the Internet is someone who starts a 'thread', hoping that someone gets hooked onto the argument. Trolls exist only to hurt others' sentiments by concealing their identity by assuming fake accounts and proxy servers.

Trolls are ruining the democratic potential of the Internet. They are turning the web into a cesspool of aggression and violence. What seeing or viewing them will do to the rest of us may be even unimaginable. Indeed, what trolls feed on is attention. Or, let's say, trolls are digital manifestation, of what may be called character assassination or gossiping.

Internet trolls are people who use anonymity persistently and many times with malicious intentions to promote and share content on Internet platforms. Internet trolls are reflections of the society's love for voyeurism. The relationship between trolls and mainstream culture is ostensibly hostile. In a marketplace for ideas, trolls are always lurking in a corner waiting to violate someone's rights and liberty.

There is overwhelming evidence that women and members of the ethnic minority groups are disproportionately victims of online abuse. More than one in five women faces harassment or is abused on the Internet, says a research by Amnesty International—a survey comprising more than 4,000 women. The study also showed that almost 60 per cent of abuses were racist, sexist, homophobic or transphobic.[1] Eight out of ten people in India face some form of online harassment, with 41 per cent of women having faced sexual harassment on the Internet, says a survey by cybersecurity solutions firm, Norton by Symantec.[2] The constant barrage of abuse, including death threats and threats of sexual violence, is silencing people, by pushing them off social media platforms and further reducing the diversity of online voices and opinions. It shows no sign of abating.

A Pew Research Center survey titled 'Online Harassment 2017' points that harassment is now a 'feature' of life online.[3] Four out of ten Americans face some form of harassment on the Internet according to the survey. Surveys show that people personally experience online abuses, with severe forms of harassment, including physical threats and stalking. Women describe online harassment as a major problem.

Women are not the only target. They target many other groups in society who are perceived to be the enemies of those who troll. Trolling is a well-calculated act, with wicked intentions of silencing others' voices. It tries to supress Dalits, women and other oppressed groups from raising their voice against perpetrators of crime against them. So, the victims are targeted in a way that they lose their confidence, and they are belittled.

It tarnishes a person's image, spoils hard-earned reputation, incites enemy, spreads fake publicity, increases visibility and encourages attention seeking behaviour. It is not easy to be a liberated woman in a patriarchal society. If they have to step out of the house, they will have to prepare themselves for potential molesters, stalkers, harassers and moral policing because such behaviour is constantly normalized. However, that isn't the end of this story. People from various sections of the society are constantly put under the radar of trollers.

The Internet raises a fundamental issue in today's society. If there is anonymity, how long can one behave responsibly? Anonymity acts like a shield that protects a person from being identified for their deeds in cyberspace. The anonymity of actions does not lead to consequences that might physically affect someone. So, would you perform something immoral because the Internet provides anonymity?

Anonymity is a serious issue on the Internet. It should not be confused with privacy issues. To that matter, secrecy and anonymity draw close parallels but not privacy, which is a highly debatable issue in our time.

Types of Trolls

Trolls stir the divisiveness that already exists in the Indian social structures. I made a list of incidents in which trolls target a particular community, such as politicians or women. This is neither exhaustive nor representative, if based on personal observation.

Celebrities

Although film stars and celebrities love to be in the news, some of them are dragged into controversy for no reason but that they had made a statement or shared photos or videos on social media. Sporting icons and film stars frequently defend their reputation, especially when the controversy is related to how they look.

Spiritual gurus such as Baba Ramdev, Sri Sri Ravi Shankar, and Jaggi Vasudev are trolled the most on the Internet. Their comments and activities are often brought under scrutiny by trollers. For example, Sadhguru Jaggi Vasudev of Isha foundation was trolled when he had tweeted an outdated video, linking it to the Swachh Bharat Abhiyan.[4] The video was actually shot in a national park in South Africa three years ago. The trolls related to this incident were brutally funny.

In another episode, Sri Sri Ravi Shankar, an Indian spiritual guru and Art of Living founder, made a statement on the Nobel Peace Prize winner Malala Yousafzai, which invited trollers to the scene. During his visit to drought-hit Latur in Maharashtra, he stated, 'That girl got the prize for nothing. The prize is a political one and has no meaning.'[5] He also added that he refused to accept the prize when the committee offered it to him. This statement was trolled widely. He also made a statement on homosexuality, where he said that it was 'a tendency' and not permanent.[6] This statement was countered by Bollywood actress Sonam Kapoor who tweeted that it is not a tendency, but something you are born with and is absolutely natural. Sri Sri's followers trolled her brutally and many asked her to apologize.

Politicians

Politicians are favoured the most by trolls. PM Narendra Modi, Arvind Kejriwal, Rahul Gandhi, Manmohan Singh, Mulayam Singh, Sushma Swaraj, Smriti Irani, the list is endless.

Narendra Modi, the prime minister of India, gets trolled, a lot. From his term as prime minister, to his statements and activities, people find every reason to troll anyone for a human error. This type of trollers find the minutest of things in their radar to jump into action! They observe everything carefully. PM Narendra Modi's international trips alone flooded social media with trolls. His comments on bringing back the Kohinoor to India or about ancient plastic surgery in India brought people together in a trolling gala. In a speech during a session of the Upper House, he made a comment against former Indian Prime Minister and Economist Manmohan Singh that he possessed a talent for getting away with a clean image even though numerous scams hit India during his term.[7] This comment also invited trolls to the level that some reminded him that if the quantum of education of all the ministers in Modi's ministry were to be added, it would not add up to that of Singh's. His demonetization policy invited criticisms. His greetings, when he starts a speech, *'Mere pyare deshvasiyon'*, alone became the basis of many trolls. And, these types of so-called criticisms are unending.

Opinion

Writers, journalists and thinkers are victimized by trollers. Nidhi Razdan, Dhanya Rajendran, Sagarika Ghose, Rana Ayyub, Barkha Dutt are examples of people subjected to opinion trolls.

Alia Bhatt and Shraddha Kapoor were trolled badly for voicing their opinions for a firecracker-free Diwali in India. They stood in favour of animals that get affected by the noise of these crackers. This was extensively trolled and some demanded them to become vegetarians and stop using makeup made from animal fat.

Shashi Tharoor, politician and author, was trolled on the social media when he tweeted in Hindi on World Hindi Day; he had made an error. When he wished all on Twitter on Mahavir Jayanti, he mistakenly shared the picture of Gautama Buddha instead of Mahavir, which instigated trolls.

Virender Sehwag, former Indian batsman, is famous for his witty comments on Twitter and often gets trolled. He was trolled when his prediction of India defeating Sri Lanka in the ICC Champions trophy did not come true. When the banning of old currency—demonetization—was announced, he tweeted for people to be patient and wait for the nation to be rescued just as Lance Naik Hanumanthappa waited for six days to get rescued at a -45 degree centigrade, 35 feet under the snow. Even though it was a well-intentioned tweet, it was dragged into controversy.

Bestselling author Chetan Bhagat is another trollers' favourite. He is trolled for even the silliest things, and in fact, whenever he makes a comment. The vilest attack was when he posted the picture of his book *One Indian Girl* and asked people who follow him on his profiles to send pictures of his book with a beautiful background. He was probably attempting to promote his book. However, social media users responded to his request with sarcasm. Many people on Facebook and Twitter reacted in a brute way, which had a photoshopped image of the cover on a toilet paper, others put up a picture of the book in a waste bin, while others put up pictures of the book being set on fire.

Indian columnist and novelist Shobhaa De often gets trolled by social media. In 2017, when the relations between the US and North Korea grew worse and when Trump addressed Kim Jong Un as the 'Rocket Man', he countered it by calling him a 'mentally deranged dotard'.[8] People rushed to find the meaning of 'dotard'. It meant an old person or a weak man. This time, De took to Twitter to make a list of desi dotards and requested people for their contributions to the list that backfired and many asked her to add herself as the first person on the list. When Indian women performed well in the cricket world cup, she tweeted a prayer to protect those women from the commercialization that the Men in Blue have fallen for. This made the fans troll her yet again. They even called her useless.[9]

Historian and writer Ramachandra Guha was a subject of social humour on Twitter, particularly for his appointment into the panel by the Supreme Court to run the affairs of the Board of Cricket Council of India, the cricket authority of the country.

The Internet has become an echo-chamber where people love to get information about things they like, while distancing themselves from other kinds of content, which are in stark contrast to their likings. Facebook and Twitter are not concerned if you have not expressed an interest in a particular content being shown to you. This is what happens over time—your news feed becomes narrower in terms of content and perspective by virtue of what your friends are sharing and what sites you visit through your social media profiles. Because reading newsfeeds of your friends is increasingly becoming the way you acquire information about the world and its issues, you are less likely to encounter the information that comes from outside your group. Thus, you remain unaware of other arguments and differing views. In today's hyper-connected world, you are all supposed to be just a click away from each other. This should be bringing you closer. However, you are finding yourselves pulled into the closed groups which is distancing you from the world.

We thus end up organizing ourselves in a largely like-minded online group, sharing information we agree with. At the same time, these isolated eco-chambers prevent us from having differing perspectives crucial for becoming well-informed citizens, tolerant of others' views. But these closed groups push you towards more extreme ways of thinking and polarizing society. They also help spread misinformation and fake news. If you spend all day receiving content from trusted friends reflecting exactly what you already believe, you are less likely to develop the healthy thought process of inspecting things before considering the veracity of a story even before you share it.

Our opinions are influenced by selective content being shared on our social media. Indeed, there is a growing tendency to silence and malign people who have a different opinion. Opinion trolls expose the level of tolerance we have in our democracy for opinions different from ours.

Clothes

Women are abused on the Internet for the clothes they wear. Priyanka Chopra dealt with trolls for wearing a knee-length dress to an event in Berlin, where she met the Prime Minister of India, Narendra Modi. She sat with legs crossed in front of the PM and shared a photo on social media herself. The Internet found her attire highly inappropriate and against the aesthetics of Indian culture, and her posture disrespectful. Nia Sharma, a TV actress, was brutally trolled for her attire as well. During the Gold Awards 2018, she walked on the red carpet in a white dress, but her fans were not impressed. They trolled her style; some comments were very inappropriate. Esha Gupta, the Bollywood actress, has been slut-shamed ever since she started posting her photoshoots on Instagram. Actresses Radhika Apte, Ameesha Patel, Taapsee Pannu, Fatima Sana Shaikh, Sonam Kapoor, Priyanka Chopra and Mahira Khan are other examples of actresses being trolled on social media.[10] Ridiculing an attire reaches a whole new level when it comes to national award winning actor Akshay Kumar. He took to social media platforms to spread a word about the auction of the navy officer's uniform which he wore in the movie *Rustom*. People trolled the movie, claiming that it is insulting for an army officer who gets the post after putting in a lot of hard work.

Body Shaming

We all have one thing in common—our body. Taboos are socially constructed in such a way that one group becomes benefactors by the negation of others. It is one potent tool developed by the Internet world. Body shaming is an attempt to disparage someone based on the shape, size and appearance of their body. The incentive behind it is to squash their confidence, spoil their peace of mind or just enjoy sadistic pleasure from the mental stress caused due to harassment. This issue is so severe that a survey by Fortis Healthcare conducted among 1,244 women across 20 cities reported that close to 50 per cent of women have experienced body shaming.[11] Now the Internet has been using body shaming as a weapon against women. Mithali Raj,[12] India's women's cricket captain was body shamed for having a patch of sweat on her top by @ashimdchoudhury. When Priyanka Chopra wore a tricoloured dupatta with a white top and denims, she was slut-shamed by a commenter for disrespecting the tricolour by wrapping it around her 'filthy body'. Parineeti Chopra, Indian actress and singer, faced a series of body-shaming trolls, where she was scrutinized for her body and was often told by her haters to eat less and slim down like other actresses. Shruti Haasan, Indian film actress and singer, was body shamed for her lip augmentation and gaining weight for her role in the movie *Behen Hogi Teri*. Actress Aneri Vajani was severely body shamed after she shared a photo of herself in a bikini. Several trolls were ruthless, petty and called her a vulture, ugly and malnourished.[13] Even wives of sportsmen are targeted. For example, Indian cricketer, Mohammed Shami's wife, Hasin Jahan, was trolled on Twitter for not wearing a hijab.[14] These are just a few cases. The Internet has plenty more.

Men are not exempt from being body shamed either. A photo captured during the India-South Africa match at the ICC Champions Trophy went viral on Twitter. It was a picture of a white woman and an Indian man sitting alongside during the match. People pointed out that the guy would never get to date the lovely foreigner because of the way he looked and the fact that he was dark skinned. Racism and casual body-shaming applies across genders.[15]

When Amy Jackson shared a selfie of her and the superstar Rajinikanth on Twitter, to which Ram Gopal Varma, the Indian film director, said that Rajnikanth has broken the notion that looks are the most important to attain stardom. This comment made the diehard fans of Rajini troll Varma.

While it seems that women are usually trolled for the way they look, it is more due to the reason that the trolls can't accept a beautiful woman to be smart and have an opinion. However, there are women like Anushka Kelkar who started the Instagram account BrownGirlGazin, where she posts portraits of women, intending to redefine beauty oustide conventional portrayals of women's bodies.

Nationality

Sania Mirza, the Indian tennis player, was always trolled for the dress she wore during tournaments. However, when she married the Pakistani cricketer, Shoaib Malik, she was further abased. Trolls questioned her love for her nation and her identity.[16] Whenever she made a comment on any issue related to India, she got trolled. Singer Adnan Sami was trolled by Twitter users in Pakistan after he tweeted on the Snapchat fiasco.[17] Sami had applied for an Indian citizenship, which he got in 2015. Bollywood star Shah Rukh Khan faced the wrath of trolls[18] in 2018 when they found one of his cousins, Noor Jahan, was a candidate in the then forthcoming general election in Pakistan. Actor Aamir Khan and his wife Kiran Rao were trolled for their patriotism, which, according to some fringe groups in India, was fake, and the actor was perceived to be anti-national.

Gender

Sushma Swaraj, the then external affairs minister, was trolled on social media on June 2018 on the matter of transfer of an officer. Journalist and author Rana Ayyub received rape and death threats online, making her a victim of 'doxing', where her personal details, including her address, were shared online. Even though neither Sushma Swaraj nor Rana Ayyub belong to different professions and are bold enough to deal with such an audience, they have been harassed in an attempt to silence their voices.

Online violence against women is essentially an extension of the society's bias against them. It targets their sexuality, objectifies them and reinforces gender stereotypes. Online violence tends to silence women or leads them to self-censor themselves because of the fear of being lashed out at. This is also why several incidents of cybercrime against women go unreported.

Alia Bhatt, one of the popular cine actors, has been always surrounded by controversies. In 2016, she said in an interview to *Cosmopolitan India* that she was not a feminist. Some trolls said she doesn't want women to be equal with men. She stirred another controversy with a new haircut (bob fringe cut), which appeared in the *Elle* magazine (2017). Look at the voyeuristic and patriarchal mindset of social media users! Social media pages were created exclusively to troll Alia Bhatt; they share content to only troll her. Women are trolled just because they are women and if they are *famous*! It is jealousy and fear of women which instigates such gender trolls.

Anatomy of Internet Trolling

Internet is no short of people who, instead of winning arguments based on reason, resort to abuse, threats and bullying to prove their point. Despite incidents occurring in different times and different places, what holds them together is the sociology unique to all of them. This type of behaviour is a measure of the progressiveness of the society by which one can see where our country stands.

Most people who relish the culture of trolling do so for self-gratification. Those with wicked mindsets try to vitiate people with whom they have differences, any kind of disparity or jealousy. Since the space proffers anonymity to users, the gossip culture works to their advantage.

Sexual minorities are trolled openly for asserting their sexuality. Those who sympathize with them are also targeted. In obliterating the voices, almost all conservative groups go hand-in-hand with no remorse. Trollers use phrases and neologisms that systematically demean sexual minorities and damage their self-confidence and pride.

Those who criticize reputed leaders have to bear the brunt of vicious online attacks. Those belonging to the right wing lead the trollers who systematically target and attack their leaders' detractors. These detractors, who are beyond questioning, often become targets of image damaging. Reputation built through years of hard work, loyalty and adherence to work ethics are spoiled just by a few messages. Foul language is used against the commenter. In doing so, the followers of leaders pay no heed to the idea that dissent is necessary for democracy.

It is true that people with high credibility are vulnerable to trolling. Some people who possess skills to use social media effectively are capable of doing the most outrageous things. They can coordinate efforts to tarnish others' images, liaison network among an endless number of social media profiles, spread messages to as many people as possible. Within seconds, a handful of people can spoil reputations built over years. The Internet comprises trolling of sorts—some make you famous or hail your victories, whereas others tarnish you for some reason which you don't even know existed.

Some ideologies are beyond any kind of public scrutiny. Particularly, right-wing ideologies seem to be a fortress. Those who venture into any form of engagement with rightist ideas become an easy bait of online trolling. You are not supposed to voice your opinions against conservative forces; some ideologies cannot be questioned; certain words cannot be used; some people cannot be quoted; some images cannot be used; some costumes are banned from use on social media. They all advocate some sort of hardcore ideology, which are supported by people who are intolerant towards any kind of democratic dialogue or dissent.

Body shaming is not a new phenomenon. In recent years, social media has taken body shaming to a new level. They prefer women with a slender body and fair complexion, and the women are told that this is what they should strive for. Anything outside that is worthless. Body shaming, or otherwise referred to as fat shaming, is an intentional act in which individuals are judged negatively, based on their physical appearance. Generally, men and women are fat shamed for being 'overweight' or not fitting into the image of 'thin and beautiful', which is perceived as the ideal. When a troll body shames a sportswoman, trollers ignore all her achievements and contributions to the nation and everything boils down to her gender, which indeed reasserts patriarchal dominion. In India, particularly, the trolling appears to have its roots in the age-old patriarchal desire to subjugate women to the whims of men.

Trolls on the Internet reflect and reproduce the psyche of the Indian society. It is a barometer for where we stand in the twenty-first century. It tells how retrogressive we are.

Trolling is an attempt to silence people who are perceived to be enemies by those who hide behind organized troll attacks. That keeps people quiet due to the fear of retribution. Those among us who never had their voice heard are being silenced in the age of social media. What instils fear in them is the organized trolling culture. Sometimes it acts as a deterrent against those who raise their voice. The ideals of democracy remain as ideals only. Even the most educated amongst us find it a merry place to use the vilest words and exhibit their divisive mindset. Trolling is not an abrupt response to a particular incident. It is also not a quick reaction to any incident steered from some unknown sources. It is clearly stage-managed, often by vicious intentions and a crooked mindset, which aims to tarnish the reputation of someone who is a hard-working and an honest citizen.

It is the product of a conservative social order. Caste, religion, language, region, class and a variety of other divisive factors contribute enormously to this sociology. Internet trolling is here to stay so long as inequality is not addressed. Education and awareness are the only feasible solutions, at present, one can think about.

Endnotes

1. https://medium.com/amnesty-insights/unsocial-media-the-real-toll-of-online-abuse-against-women-37134ddab3f4, accessed on 15 May 2019.
2. Yuthika Bhargava, '8 Out of 10 Indians Have Faced Online Harassment', *The Hindu*, 5 October 2017, https://www.thehindu.com/news/national/8-out-of-10-indians-have-faced-online-harassment/article19798215.ece, last accessed on 15 May 2019.
3. https://www.pewinternet.org/2017/07/11/online-harassment-2017/, accessed on 15 May 2019.
4. The video was three years old and was credited to Narendra Modi's Swachch Bharat Abhiyan. Video was about an elephant putting empty cans in recycling bin. He tweeted that Swachch Bharat was caught on and he congratulated the prime minister.
5. Express Web Desk, 'Sri Sri Ravi Shankar, Nobel Prize and the Art of Getting Trolled', *The Indian Express*, 23 July 2016, https://indianexpress.com/article/trending/trending-in-india/sri-sri-ravi-shankar-nobel-prize-and-the-art-of-getting-trolled-2782705/, last accessed 24 May 2019.
6. Aranya Shankar, 'Being Homosexual Is "a Tendency"', Says Sri Sri Ravi Shankar at JNU Event', *The Indian Express*, 14 November 2017, https://indianexpress.com/article/cities/delhi/being-homosexual-a-tendency-says-sri-sri-ravi-shankar-at-jnu-event-4936309/, last accessed 24 May 2019.

7. FE Online, 'PM Narendra Modi Targets Manmohan Singh, Gets Trolled on Twitter', *Financial Express*, 9 February 2017, https://www.financialexpress.com/india-news/pm-narendra-modi-targets-manmohan-singh-gets-trolled-on-twitter/544334/, last accessed 24 May 2019.
8. Austin Ramzy, 'Kim Jong-un Called Trump a "Dotard". What Does That Even Mean?' *The New York Times*, 22 September 2017, https://www.nytimes.com/2017/09/22/world/asia/trump-north-korea-dotard.html, last accessed 24 May 2019.
9. Meghna Nijhawan, 'Author Shobhaa De Trolled After Tweet On Women's Cricket Team', *NDTV*, 2 August 2017, https://www.ndtv.com/offbeat/author-shobhaa-de-trolled-after-tweet-on-womens-cricket-team-1732671, last accessed 24 May 2019; DNA Web Team, 'Shobhaa De Brutally Trolled After Praying For Women Cricketers To Not Get "Ruined" Like Their Male Counterparts', *DNA*, 2 August 2019, https://www.dnaindia.com/cricket/report-shobhaa-de-brutally-trolled-after-praying-for-women-cricketers-to-not-get-ruined-like-their-male-counterparts-2520275, last accessed 24 May 2019.
10. 'Bollywood Actresses Who Got Trolled for Their Outfit Choices', *Times of India*, https://timesofindia.indiatimes.com/entertainment/hindi/bollywood/photo-features/bollywood-actresses-who-got-trolled-for-their-outfit-choices/photostory/59123812.cms, last accessed 24 May 2019.
11. India CSR Network, 4 June 2019, '50% Women Body Shamed At Least Once in Life: Report', https://indiacsr.in/50-women-body-shamed-at-least-once-in-life-report/,
12. Timesofindia.com, 'MithaliRaj's Epic Reply To Twitter User Who Tried To Body-shame Her', *Time of India*, 22 August 2017, http://timesofindia.indiatimes.com/articleshow/60170129.cms?utm_source=contentofinterest&utm_medium=text&utm_campaign=cppst, last accessed 24 May 2019.
13. Sonali Singh, '"Beyhadh" Actress Poses In Lingerie, Receives Severe Backlash on Instagram for Being Too Skinny', 20 June 2017, https://www.indiatimes.com/entertainment/celebs/beyhadh-actress-poses-in-lingerie-receives-severe-backlash-on-instagram-for-being-too-skinny-324201.html, last accessed 24 May 2019.
14. https://www.deccanchronicle.com/sports/cricket/180817/mohammed-

shamis-wife-hasin-jahan-trolled-again-on-twitter-for-not-wearing-hijab.htm
15. Pooja Salvi, 'When Boys Get Body-shamed', *The Asian Age*, 14 June 2017, http://www.asianage.com/life/more-features/140617/when-boys-get-body-shamed.html
16. Anurag Verma, '"My Independence Day Is August 15": Sania Mirza Shuts Down Troll Who Questioned Her Nationality', News18,15 August 2018, https://www.news18.com/news/buzz/my-independence-day-is-august-15-sania-mirza-shuts-down-a-troll-who-questioned-her-nationality-1845075.html,last accessed 24 May 2019.
17. A former employee of Snapchat alleged that its CEO Evan Spiegel had once called India 'too poor' for the social media app. Indians were outraged over his comment and the hashtag #uninstallsnapchattrended. Along with them, singer Adnan Sami joined in the social media outrage and tweeted he 'just uninstalled Snapchat'. https://indianexpress.com/article/trending/trending-in-india/adnan-sami-snapchat-uninstall-tweet-for-india-leads-to-pakistani-twitterati-abusing-him-4616250/, last accessed 24 May 2019.
18. Troll Slayer, 'Trolling SRK Because His Cousin Is Contesting Elections in Pakistan Is Reprehensible', *Times of India*, 11 June 2018, http://timesofindia.indiatimes.com/articleshow/64538540.cms?utm_source=contentofinterest&utm_medium=text&utm_campaign=cppst, last accessed 24 May 2019.

6

Sexism on the Internet

Radhika, a schoolteacher in Tamil Nadu, is an avid social media user. She frequently shares content related to subjects of her interests such as politics and fashion. Although she enjoyed her visibility (5k followers and 4k friends on Facebook), strangers would message her asking for her number, and when she would decline, they would abuse her, send threats of rape and share obscene content. It seems that just by speaking up, a woman, in the virtual world, is 'asking' for the abuses. Initially, she would block them. However, as their number increased, she felt that the Internet has become highly inhospitable for women like her.

This is not a stray incident. A majority of women who use social media for various purposes encounter similar men, who, all of a sudden, start sending friend requests or try to become friends. They initially present themselves as gentle people, who are progressive, open-minded and liberal. However, their behaviour soon turns inhospitable and unbearable when women refuse to share their contact number, email IDs or decline their requests. The question is, how long can someone maintain silence or avoid them? This is really an attempt to silence women—if you want to silence women, use their gender against them. This is often a well-calculated move.

The Internet has reinvented the sexist social structures that exist outside it. This type of problems has been well documented in books such as *Cybersexism* by Laurie Penny and *I Am Not a Slut* by Leora Tanenbaum. Authors of these works have investigated into the attempts to silence women on the Internet. Many things go unreported for fear of the stigma, and those being aired in public grapple with the sheer wanton vandalism. Abusive language, sexist comments, threats—both physical or sexual—posting private information, including contact details (doxing), racist language, publishing intimate images without consent; the sort of violence women are subjected to is such that the Internet has become too hostile a place for them.

When a woman steps out into the streets, gender-based harassment appears to be a part of their everyday life. All because she is in a public space—a space that supposedly is not meant to be hers. Gender-based abuse, harassment and stereotyping women is as much a part of the online world as it is in the real world. Sadly, it mirrors the real world.[1] Gender stereotyping on the Internet pinpoints class, gender, race, ethnicity and communal reconfiguration in India.[2]

Doxing

Revealing personal or identifying information about someone without their consent is doxing. It includes information such as a home address, office address, real name, children's names, phone numbers and email addresses. A violation of a person's privacy, which is the aim of doxing, is done to cause distress and panic.[3] Recently, Neha Dixit's, an independent journalist, number and home address was circulated on social media. She was even intimidated by someone who had posted her private family photos on social media. She also received rape and death threats for merely doing her job.[4] Many women have reported that their personal information has being exposed on social media by fake accounts with an intention of slut-shaming them. Information shared online is easily used against someone. This type of harassment is an offshoot of the patriarchal idea that those women whose information are being aired in the public are 'bad'. So their reputation can be damaged, their confidence can be destroyed and, most likely, they would face sexual harassment of the sort that they might have encountered at a lonely corner of a street.

Hindutva Misogyny

Soon after student activist Gurmehar Kaur made a statement on social media against the violence perpetrated by Akhil Bharatiya Vidyarthi Parishad (ABVP) in the Delhi University campus, she was trolled mercilessly for an earlier #ProfileForPeace social media campaign, in which she had said, 'Pakistan did not kill my father, war did.'

The vitriolic Hindutva troll army turned on Sushma Swaraj, former Supreme Court lawyer and senior leader of the Bharatiya Janata Party (BJP), because she was involved in the transfer of a passport official, who apparently harassed an inter-faith couple. The trolls abused her saying that she was 'better off dead', she has an 'Islamic kidney', she was taken in by the 'sickular, libtard' ideas, and that she was 'good for nothing'.[5] When the news of the murder of Gauri Lankesh, who had vociferously spoken and written against the rise of communalism and right-wing politics and supported the causes of Dalits, women and transgender,[6] by unidentified assailants at her home in Bengaluru came to light, many took to Twitter and welcomed the news with great happiness.

Sagarika Ghose and Meena Kandasamy were also trolled for their opinion by conservative forces. Women are also trolled on the basis of their nationality, often connecting them with Pakistan.

Right-wing groups subjugate women, who, according to them, are non-conforming. More importantly, they troll those women who express an opinion. Even worse is the experience of women who have an opinion against right-wing politicians. The only thing that invites vicious cyberattack is that they commented against Narendra Modi, BJP or RSS or anything which, according to Sangh Parivar forces, is unquestionable. In such a situation, the attack is primarily on their gender.

Religious Trolls

Irfan Pathan's wife Safa Baig was trolled for exposing her hands and applying nail polish. Indian tennis star Sania Mirza has always been a subject of trolls, be it Hindutva or Islamists forces, ever since she tied the knot with Pakistan cricketer Shoaib Malik.

There is a pattern of trolls that seems to be predictable. Women are trolled in a predictable way because they love liberty and speak about it on social media platforms. They are trolled for marrying outside their religion or even when they are found talking to men. Women in India have to adhere to many dos and don'ts unlike men. It is so powerful that even for some crimes committed by men against them, the onus falls on the women and not the actual perpetrators. Hence, there are cases where the society's views on women being raped are because they are attractive, they are teased because they are not dressed properly or they are found loitering. The patriarchal mindset has confined women to live within the set norms wound by a 'sacred thread', breaking which means they will be declared as outcast and ostracized by society. Those who troll are so much indoctrinated by some ideology or belief system that they believe they have a right to correct women by using abusive language in cyberspace. So, there is a predictable pattern in cybertrolls. After several years, such patterns don't seem to change. Some women are tagged as Jihadis, sex slaves, etc., because of their religious identity. Sometimes, they wake up to being called an anti-Indian and spy, and face rape threats. The most predictable one is when women writers and journalists are asked to leave the country and go to Pakistan or are tagged as prostitutes and sluts, who compromise their womanhood for money.

Male Gaze

We believe that the Internet is a gender-neutral place. Actually, we believe that it has no gender. But this need not be true. Feminist theories assume that there are concerted efforts in art and literature to portray women as sexual objects. In doing so, women and their world are seen through the 'lenses of patriarchy'. Having said that, their body, sexuality, liberty and all other aspects of women and their lives are seen from a masculine, heterosexual perspective. This theory analyzes how the Internet has also become a place where the male point of view about what it means to be a woman holds significance. There are several websites, including chat rooms and dating sites, where women are shown as commodities and sex objects. On dating websites, women are flooded with messages, often obscene. Most of the times, users are anonymous. The comments posted on the profiles of not only film stars and sport personalities but also ordinary women, whose images are used without consent, are disturbing. On YouTube, songs by women are the entrenched enclaves of harassment.

Shrinking Women's Space

Women who have an opinion have a 'poor' audience on the Internet. They have fewer followers and people pay no heed to their comments and posts. No one listens to their stories seriously, and they do not even have an audience among their own gender. Neither sympathy nor solidarity is received from people who share similar interests with these women. This is astonishing as the Internet was supposed to be a place where those without a voice could have their voices heard.

Empirical surveys say that women are poorly represented as users on the Internet and they do not have a good audience. This issue was particularly highlighted in the writings of Pippa Norris, a comparative political scientist at Harvard University, who called it a digital divide.[7]

Under representation of women on the Internet and their shrinking audience has caught the attention of many writers who used interesting terminology to pinpoint this sordid state. This issue was addressed as Digital Diva by Alexis Gelber.[8] P.J. DiMaggio and E. Hargittai have used the term Digital Inequality.[9]

'Gendering the digital divide', was the phrase used by another study,[10] whereas some others use 'gendering the Internet'.[11]

Some studies provide evidence that the Internet is a gendered, classist and racist space.[12] That said, the Internet is growing into a place, which is in no way different from the real world, trodden with gender violence, caste, class and racial violence. The same problems have been reinstated in cyberspace.

Among them, one cannot avoid Pippa Norris, who says that the digital divide is creating social inequalities ever more complex. She upheld the view that access to digital technology is an important pre-requisite for a good career, educational opportunities, economic success, accessing social networks, personal advancement and opportunities for civic engagement. The underclass of information poor is likely to be increasingly marginalized in societies where basic Internet skills are important criteria for social mobility. Being a user doesn't add any value to your life. You are more likely to remain an information poor. What you need is the ability to understand social media in critical scenarios. Or say, you have basic knowledge of the aesthetics of social media. That said, you know how to distinguish between good news and bad news trending on social media, what trolling is, what fake news is and what an Internet scam is. In India, the assumption that the access to the Internet and computers are significantly a political choice in a taboo-riddled social structure holds significance.

Ultra Violet[13] is a feminist blog platform where you can find articles on issues such as identity and violence against women. They address issues that do not get adequate attention from the traditional media. Ultra Violet[14] is an abode for looking at the ways in which young women are challenging, opposing, negotiating and transforming unequal power structures. Unfortunately, they do not have enough followers. Such platforms come up like islands, forming their own community. The Internet is increasingly becoming enclaves of like-minded people where they follow what they like and unfollow what they dislike. They post and share things they like, whereas by default, distance themselves from everything they don't want to see. This is the destiny of almost all marginalized groups on the Internet. They exist in the micro public space of a few people where their messages are not disseminated to a wider audience.

Endnotes

1. Danish Raza, 'Study Confirms Abuse Of Indian Women Online: Here's How To Stop It', *Firstpost*, 15 April 2013, https://www.firstpost.com/tech/news-analysis/study-confirms-abuse-of-indian-women-online-heres-how-to-stop-it-2-3623665.html, accessed on 30 April 2013.
2. See, RichaKaulPadte, 'Street Talking', *Caravan*, 1 July 2013, https://caravanmagazine.in/perspectives/street-talking, accessed on 24 October 2013; Kavitha Sharma, 'Cyberspace Violence', *The Hindu*, 11 May 2013, https://www.thehindu.com/opinion/columns/Kalpana_Sharma/cyberspace-violence/article4698886.ece, accessed on 24 October 2013; SairaKurup, 'Online Abuse Of Women Increasing In India', *Times of India*, 5 May 2013, https://timesofindia.indiatimes.com/social/Online-abuse-of-women-increasing-in-India/articleshow/19890821.cms, accessed 24 October 2013; Richa Kaul Padte and Shehla Rashid Shora, '#Misogynyalert: The Grievous Threat To Women Online', *The Sunday Guardian*, 20 April 2013, http://www.sunday-guardian.com/artbeat/misogynyalert-the-grievous-threat-to-women-online, accessed on 24 October 2013; Mahima Kaul, 'The Big Issue For Indian Web Users', *UNCUT*, 15 April 2013, http://uncut.indexoncensorship.org, accessed 24 October 2013; Asha Mahadevan, 'Highway Harassment', *Mid-Day*, 12 March 2013, https://archive.mid-day.com/news/2013/mar/120313-highway-harassment.htm, accessed 24 October 2013; Jim, 'Is Online Misogyny More Prevalent Now Or Just More Evident?' *Leaderwest*, 7 March 2013, http://leaderswest.com/2013/03/07/is-online-misogyny-more-prevalent-now-or-just-more-evident/, accessed 24 October 2013; A study titled 'Women and the Web' by Intel reported that one in five women in India believe that the Internet is not appropriate for them. (Emma Barnett, 'Fifth of Women in India Think Internet Use Is "Inappropriate"', 16 January 2013, http://newindianexpress.com, viewed on 30 April 2013). A series of essays by the Internet Democracy Project (IDP), shows the growing public concern over harassers. (http://internetdemocracy.in/search/online+misogyny/, accessed on 23 October 2013). With new technologies, the enormous potential to diffuse information, to overturn modernist notions of male sovereignty and to improve women's everyday lives seem to have been misused. (EileenGreen and Alison Adam, *Virtual Gender: Technology,*

Consumption and Identity Matters, New York: Routledge, 2001). Internet Democracy Project (IDP), a Delhi-based NGO, published a disturbing account of crude sexual and violent social media posts (www.internetdemocracy.in). A blog post by Shehla Rashid recounted that hundreds and thousands of online verbal abuses remain unnoticed (Shehla Rashid, 'Gendered Abuse Online', Internet Democracy Project, 22 March 2013, http://www.internetdemocracy.in/2013/03/22/eroticsindia-blog-8/, accessed 12 March 2013).

3. Andrew Coates, The Conversation, 'Doxxing, Swatting and the New Trends in Online Harassment', *Scroll.in*, 23 April 2015, https://scroll.in/article/722509/doxxing-swatting-and-the-new-trends-in-online-harassment, last accessed 24 May 2019.
4. Garvita Khybri, 'How Threats on Twitter Manifest In Real Life: Indian Troll Tales', *The Quint*, 28 August 2018, https://www.thequint.com/neon/gender/trolling-women-journalists-rana-ayyub,last accessed 24 May 2019.
5. NH Web Desk, 'Herald View: Hindutva Trolls and Sushma Swaraj; When Will BJP Learn?', *National Herald*, 29 June 2018, https://www.nationalheraldindia.com/editorial/hindutva-trolls-and-sushma-swaraj-when-will-bjp-learn, last accessed 24 May 2019.
6. 'How Trolls Shamelessly Celebrated Murder of Gauri Lankesh', *Oneindia*, 7 September 2017, https://www.oneindia.com/india/how-trolls-shamelessly-celebrated-murder-gauri-lankesh-2538203.html, last accessed 24 May 2019.
7. Pippa Norris, *A Virtuous Circle: Political Communications In Post-industrial Democracies*, Cambridge: Cambridge University Press, 2000
8. Alexis Gelber, 'Digital Divas: Women, Politics and the Social Network', Discussion Paper Series, Joan Shorenstein Center on the Press, Politics and Public Policy,#D-63, June 2011.
9. P.J. DiMaggio and E. Hargittai, 'From the "Digital Divide" to "Digital Inequality": Studying Internet Use as Penetration Increases'. *Working Paper 19*, Princeton, N.J.: Center for Arts and Cultural Policy Studies, Woodrow Wilson School, Princeton University, 2001.
10. Tracy Kennedy, Barry Wellman and Kristine Klement, 'Gendering the Digital Divide', *IT&SOCIETY*, 1 (5): 72–96, Summer 2003.
11. Liesbet van Zoonen, 'Gendering the Internet Claims, Controversies and Cultures', *European Journal of Communication*, 17(1): 5–23, 2002.

12. Lori Kendal, 'Meaning and Identity in Cyberspace: The Performance of Gender, Class and Race Online', *Jai Press Inc*, 21(2): 129–53,1998.
13. http://ultraviolet.in/
14. http://youngfeminists.wordpress.com

7

Internet Narcissists

The concept of narcissism dates back to thousands of years, when Ovid wrote the legend of Narcissus. Narcissus was a handsome young man. Upon seeing his reflection in a lake, he fell in love with his own image and could not leave its bank. Narcissistic personality disorder has its earliest roots in this ancient Greek mythology.

The psychoanalyst Sigmund Freud, through his work on the ego, popularized the concept. In 1914, he published the famous paper *On Narcissism: An Introduction*. He suggested that narcissism is connected to whether one's libido (energy that is part of each person's survival instincts) is directed inwards, towards one's self, or outwards, towards others. This work became the entry point for many others developing theories on narcissism.

Narcissism lies on a spectre of healthy to pathological conditions. Healthy narcissism represents healthy self-love and self-confidence that is based on real achievement. It is the ability to overcome faults and weaknesses and derive the support needed from social bonds.

However, narcissism becomes a problem when the individual becomes preoccupied with the self, expecting excessive admiration and approval from others while showing disregard for other people's sensitivities. This becomes a pathological behaviour.

Narcissists portray a highly imposing and pretentious self-image to the world. They present everything in such a way that it grabs the maximum attention of the people. But behind this mask forms the very weak and vulnerable inner sense of self. This is the mask of narcissism. Behind it, they want to cover up a deep sense of insecurity and fragile self-esteem. They can easily be wounded, have low confidence, and are unable to stand up to things.

Some writings have labelled narcissism a modern epidemic. The tremendous changes brought about by industrialism and technology have profound implications on individuals and their connection with society. The past few decades have witnessed tremendous changes in society where there was a shift from the society's commitment from a collective to more solid focus on individuals. As our society is increasingly becoming individualistic, narcissistic traits are here to stay. The rise in technology and the development of hugely popular social networking sites such as Facebook, Twitter, etc., has further changed the way we spend our free time and communicate with the world and also how we present ourselves to the rest of the world. Here, narcissism and social media draw close parallels.

The emergence of the self-expressive narcissism is the most pronounced in the online world. We are allowed to express ourselves, but what does it mean when someone keeps sharing things about themselves? While it is understandable to let the world know you are doing something, it does not make sense when your social media newsfeeds are only about *you*. This self-expressive narcissism is spoiling the real character of the

Internet. Why is it that people tell more about themselves on Facebook and Instagram, where accountability is lacking? In doing so, people are too bothered whether others like it or are even interested. The damage narcissism brings can be quite amorphous and ill-defined.

Narcissism on Social Media

We all like to be appreciated for things that hold meaning for us and we all love to share this information among friends, relatives and acquaintances. But on social media, this sharing has become excessive and unreasonable. Friends post selfies of themselves at exotic destinations and humblebrag about their career and achievements. Often, the intention is to make viewers feel that their life does not measure up to theirs. Although one is aware of how life online and offline differs, it's still easy to fall into the trap of other people's perfect social media profiles that convince you that your life is somehow falling short of such events.

Let us assess Internet narcissism from different angles. Some people on social media have an opinion on almost everything under the sky and use social media as a tool to let the world know their points of view. They watch movies and write reviews, read books and belittle the author; travel a lot and update photos and find pleasure in making others jealous; eat at a luxury restaurant, and post about the food and ambience; unnecessarily judge the government and comment on other political groups. They comment on everything—from a coffin to rocket science—the Internet has become a platform for narcissistic people.

There is another very interesting twist in the narcissistic use of digital technology. From posting deceptive selfies, in which an average-looking individual appears charming, to sorting social media feeds such that one is seen doing amazing things, social media has unleashed our self-obsessed and ego-centric nature.

Curiously enough, some people present themselves as if they are always doing amazing things and advertising themselves as demi gods, who have solutions to almost everything. On the Internet, they pretend to be Good Samaritans, saviours, prophets and emancipators. Some social media users feel the need to become 'mini-celebrities', watched and admired by others on a daily basis. They are the Internet saints who live alongside us. There are too many updates about their activities, which portray the stories of vanity and solipsism on the Internet. Millions of photographs are uploaded on Instagram every day, which gets millions of likes every day. Millions of people publish details of their private lives on the Internet. The social media is turning relatively modest Homo sapiens into a pack of 'publicity-hungry sapiens'. The trouble with this aspect of the Internet is that nearly everyone presents an unrealistic portrait of themselves. Just as people select the most attractive photos of themselves to use as profile pictures, they tend to populate their newsfeeds with the most attractive version of update about themselves.

Some people on social media are experts in precipitating positive reactions to their update within just a few hours of posting them. Their strategy is called exchange economy, where people post insincere responses on like-minded updates of others. It is sort of a give and take approach. The yearning for recognition is so pronounced that it has led people to pimp themselves out to the world by the principle 'like and share my post, and I will like yours'.

Social media responses to those narcissistic updates are clear signs of the direction in which the Internet is moving. Facebook, Twitter, Instagram and others have become places where you cannot easily buy 'likes' and followers. You cannot quickly build an audience here. You need to come up with something sensational and emotionally appealing to an audience. It has already cut into our psyche that shows where our society stands now.

The ability to disguise yourself as whoever you want to be and perceived as by your followers makes social media the favourite for some users. It is the perfect place to create a fake image and get attention for it.

Look at the irony of narcissism. Unhappy people can create an illusion of being happy; those who are lonely can create an illusion of having a huge network of friends; and 'bad' parents can create an illusion of being 'good' parents. Social media is a place where users can create many illusions. In the larger context, a user on social media has a view of the world, which is almost determined by the behaviour and opinions of the individuals they are connected to on the Internet. However, that is not the real world. They are seeing the world only through others' opinions and views.

Abdul Khadar, a professor in a government college, would update every trivial detail about himself on his social media. His followers, unaware of the real story, kept congratulating him every time he updated about his professional achievements—selection for the Fulbright Fellowship, or being recognized for the Research Author Award by a reputed publisher, or when the media talked about his success. A photo that showed Abdul Khadar receiving the award from a legendary Bollywood actor made him quite the inspiration among his followers.

However, a few days later, local dailies reported that a college professor had morphed pictures of a film award function and posted pictures of himself receiving a literary award from a celebrity. The news further reported that the professor had falsely claimed to be the winner of the Research Author Award, and that a north India-based publisher was planning to issue a legal notice against him for misusing their name. Soon after, his name had disappeared from Facebook and he had left WhatsApp as well; he reportedly deactivated all his social media profiles.

The social media shows our double standards and vile narcissism as well. You may never know, a person speaking against crimes on social media may be a criminal herself or himself. Who is to know if an angry rant against the growing economic inequality is from someone who has fleeced the people to become a millionaire himself!

At the end of the day, all of them talk about simplicity, kindness and compassion on the Internet, but in their intimate and private life, they are far from it.

Types of Narcissists on the Internet

Some narcissists are know-it-all's. These type of people are always eager to give their opinion on almost everything. They are active on Twitter where they tweet on and retweet political events. They give unsolicited advice and comments and believe that they know more than anyone else, even though the topic of conversation is not their cup of tea.

1. **Grandiose narcissists**

   This type is more clearly demonstrated as a familiar kind of narcissism. They see themselves as a superior breed and more influential and powerful than everyone else you know. They over-share details of their achievements, relationship, knowledge, access to power and networks. They tout their own accomplishments, foreign visits on invitation, exaggerate their importance and elicit your envy or admiration. They believe they are destined for great things.

2. **Seductive narcissist**
   A seductive narcissist, unlike the other types of extreme narcissists, manipulates you by making you feel good about yourself. Their intention is clear. They want you to behave the same way they do. At first, they will admire and even idealize you. In doing so, they intend to make you feel the same way about the other narcissist. They want your support and admiration and will flatter you in order to be admired. Once their purpose is met and you are of no further use to them, they cut you off.

3. **Bullying narcissists**
   A bullying narcissist builds his confidence up by humiliating others. They are always right. They are more brutal about the way they assert their superiority. They often rely on contempt to make others feel inferior, proving themselves winners in the process. They will belittle and mock anyone, and when they need something from you, they may even start threatening you. In the end, they will make you doubt yourself and your worth as a human being with any potential.

4. **Vindictive narcissists**
   Vindictive narcissists try to destroy you. It is perhaps because you may have challenged their status in some way you would not even realize. As a result, they need to prove that you are the ultimate loser by destroying your self-worth. They will trash you on social media, use difficult words, write long messages that are not always comprehensible and offend you in front of your friends and family.

The Impact

Narcissism is vividly portrayed on social media. Social media has become a place to vent narcissistic expressions to terrorize the society. The fact is that, narcissists have always existed in the society. Social media has just given them an extra tool to terrorize people.

Studies[1] have shown the link between the rise in narcissism and the usage of social media. Neither is expecting quick and positive reactions on your posts in the form of likes and comments normal behaviour, nor is assessing friendships and relationships through the number of likes and shares on your post a normal way of social communication. Having said that, social media has become a place to vent all sorts of abnormal behaviour.

Here are some tacts that narcissists on social media use to exploit, terrorize and destroy people.

Love Trap

Thanks to social media, one can enjoy multiple affairs in covert and insidious ways. Narcissists tend to be insatiable in their attention-grabbing activities and wish to create harems of people who adore them. It is easy to convince anyone by false claims and fabulous promises. Anonymity works like a shield as no one knows your true identity. Ordinary people are easily convinced by narcissists to follow them in cyberspace. Narcissists on social media flirt with numerous people by sending messages on Facebook and WhatsApp. Their comments are inappropriate; they create followers by using several social media accounts to portray themselves as endearing people.

There isn't any specific way in which narcissistic people would lay their love trap. But they have a clear target, including teenagers who are new to social media, in their mind. They are likely to choose middle-aged housewives, who are perceived to be 'unsatisfied', widows and single women. Women who are obsessed with their looks and those who always appear with a lot of make-up in selfies also become their target. These predators tend to not go after young independent women who are bold enough to take a stand against those who supposedly harass them. These predators instead try to troll such women, and their weapon is character assassination.

The fact is that they contact them with well-prepared scripts. They want admiration, recognition and love. These serve as motivation. Their aim is to enjoy the pleasure derived from the attention people give them. They go to any extent to get narcissistic pleasure.

Social Listening

Social listening means the monitoring of users' social media activities. It is a part of social analytics where brands build their customer base for selling their products. But that is not all. They closely watch your behaviour in the cyberspace to find out your tastes and interests so that they can attract you to their profile. This sort of behaviour on social media is a clear indication of the opportunities it provides people with a criminal bent of mind to intrude into the lives of innocent people and create trouble. The data leaked by way of social listening can be used to build a huge list of fan following for narcissists on social media so that they can be misused.

Internet narcissists are like trained spies who use social media accounts to know all the details of any number of people so that they can use it against you. It is even possible for a complete stranger to find out the identity of other people by looking through your comments, photos, shares and tagged posts. This type of investigative digs is crucial for the narcissists to infiltrate your vulnerabilities.

It allows them to assess whether you would be a viable target for their pity ploys and mind games. They communicate with any number of people on the Internet to build trust despite the fact that trust has to be built organically and cannot be earned blindly.

Stalking and Harassing

Narcissists are always self-obsessed with people that intrigue them, such as exes or celebrities. These days the entire cyberspace is set up to nourish and amplify these self-obsessions, and this is where things can get tricky.

Stalking in cyberspace is a criminal practice, where a narcissist uses the Internet to systematically harass or threaten someone. It occurs in conjunction with the more traditional form of stalking, in which the offender harasses the victim offline.

Narcissistic ex-friends, partners or colleagues or employees never let you go, even after the ending of a relationship. Even if one of their accounts is blocked, they will follow them using other fake accounts. Anonymous e-mails, impersonation and public insults thereby haunt you frequently from all fake accounts. They troll their ex-girlfriends or ex-colleagues. All this is a way to make others feel unsafe. It is a way to cast narcissistic assaults on another person's agency in the cyberspace. The feeling of someone always watching and monopolizing your very presence somewhere is terrifying. It creates a sense of violation that is rarely prosecuted in the realm of the law. That is what the Internet narcissists just want it to be.

Self-aggrandizement

Impression management is vital to a narcissist. Grandiose narcissists have an innate sense of superiority, and they are more likely to be found glorifying themselves on social media as opposed to other narcissists with lower self-esteem. They are able to use social media to disguise their true nature under a charitable mask and build fan clubs and blind followers who encourage their toxic behaviour.

For a malignant narcissist, the activities go beyond self-absorbed selfies. It ventures into a complete lack of self-awareness and empathy for others. However, not all people with numerous selfies and thousands of friends will meet the criteria of a fully-grown narcissism nor should these be the only criteria of judging a narcissist. A malignant narcissist uses social media not just as a hunting ground but also as a platform to show and practise his or her grandiose and assumed superiority. They engage in lengthy monologues about things even if it bears no resemblance to their actual behaviour.

Grandiose narcissists roam free to eulogise themselves while cyber bullying others. They are the instigators of manipulative arguments on discussion forums, and debate clubs on cyberspace—the one who carries out multiple counts for character assassinations in a variety of fields from family to head of state. They include seemingly philanthropic activities and leadership qualities they claim to possess in their posts. They try to attack others for lacking what they claim to possess. They compete with others when it comes to putting up a status rather than focusing on what their message implies. Indeed, they are the faceless, superficial models and somatic bodybuilders who create a self-built image based on the charity of the audience.

Their views and opinions may be different, but what all of them share is an excessive sense of entitlement with little to no regard for the rights or needs of other people in the cyberspace. When the self-aggrandizing narcissist speaks, they demand your attention. They demand that they be accommodated into your lives. They feel entitled to your time and resources.

Endnote

1. https://www.psychologytoday.com/us/blog/compassion-matters/201211/is-social-media-blame-the-rise-in-narcissism, https://www.sciencedaily.com/releases/2018/11/181109112655.htm, lastaccessed 17 May 2019.

8

Automation and Relationships

Shalini Sreedhar is the only child in a family of three. She felt devoid of companionship and lonely, and compared herself to an orphan. Her mother, a doctor, and father, a busy business executive, return late from work and Shalini had to spend most of her day after school by herself. Like any other normal child of her age, she missed her parents terribly and looked for solace. However, when she found Facebook, she realized it was a great way to fill the void left by her parents. She no longer missed her parents; and did not even crave for their love and affection.

Like Shalini, many children belonging to middle-class families are finding it more comforting to spend time on Facebook than with their parents. While parents are chasing money, profession and fame, the children are left to deal with an emotional void. Facebook and WhatsApp have filled this void created by the absence of their parents. In the labyrinth of silence and the wilderness of loneliness inside the walls of their homes, they look for a change. It is due to these gaps left by our social fabric that their minds wander towards social media.

Meet Athira, a young girl who hoodwinks her mother. Her mother had requested her to study, but she had no idea that Athira had hidden her phone inside her book, and she was pretending to study. Actually, she was Facebooking on her mobile. Although this is a distraction, this is how young people create a new social space. This happens in every family.

Social media is a place where crimes are increasingly being reported and those perpetrators go unidentified because of the anonymity the Internet provides. They learn from social media how to commit the crimes that are otherwise forbidden.

As I had mentioned earlier, I used to meet people from different walks of life to understand how social media operates. Among them was a teenager as well, whose insight helped me understand further how social media pollutes the society and distracts people from their purposes. Let me share one of her stories. She told me that she had met an engineering student on Facebook and they shared an intimate relationship. I asked her, 'Don't you find anything wrong with that relationship?' She said, 'No, older people, like you, may not be in a position to understand such a relationship. However, I can, as I belong to this generation.' Indeed, Facebook has given her a chance to be in a non-conforming relationship.

These are just random scenarios I have come across, which pinpoint an emerging social awareness in India. Face-to-face interactions meant that one could understand the other person's emotions. Now, social media has replaced such interactions with emoticons and acronyms where people communicate their emotions through them. Teenagers who are frequent to social media sites are more likely to develop poor social skills and understanding ability. This can affect democracy and social scenarios of the future of India. The new generation of social media Indians are cool and smart; they care little about things that previous generations held in reverence.

We have reached a point where we cannot live without digital technology. Now, separating ourselves from our smartphones seems impossible and not having WiFi or mobile data seems like the end of the world. If our phones die, we feel helpless and are unable to function. If the Internet is not accessible, we become nervous. It is difficult for people to imagine a situation without digital technologies of connection.

Uninstalling social media apps from the mobile phone and going to the nearest coffee shop with a close friend seems like an old idea. Calling your mom on the phone instead of sending a text on WhatsApp is a thing in the past. Writing a letter and sending it via the postal services to someone that you love makes people laugh at you. Turning off YouTube and going outside to watch the beauty of nature has become a primitive idea. Getting a group of people together to loiter in the streets or playing some kind of sports seems old-fashioned. Above all, reading a book makes people call you an impostor. Technology, indeed, has brought profound changes where what was considered the good old ways of doing things seems antique for the data generation.

We are becoming slaves to our technology. It affects our happiness and our sense of stability, and we tend to talk and understand each other lesser. We live on and through our digital screens. We are always plugged into some kind of gadget. This also means we are increasingly distracted.

Our smartphones are constantly buzzing and our lives are interrupted by notifications. They keep begging us to pay attention to something or the other. They make us drop everything to check what the notification could possibly be about even if it isn't something important. Technology has become our closest friend. It has become a person we will do anything for.

People feel lonely, lost and stripped of their identity without technology and their smartphones. Their holidays now must include a communications device. A trip without photos that they can immediately upload on Facebook or Instagram doesn't qualify as a trip.

Some years ago, the Internet was a specific tool designed for the academia in select nations. Now that has become a medium for people across the globe. Never before has any technology found such an intimate relationship with users in such a short period.

This remarkable shift, for sure, has had many benefits to human life. But the ability of the mind to concentrate on things diminishes with prolonged exposure to the Internet. In essence, technology is making people addicted. People are always posting only one side of their stories, just as one is posting selfies, which exhibit unrealistic beauty standards. It cannot be trusted.

The Internet is a friend who demands faith and submission. There are many people who submit to it unconditionally. A new world is re-emerging in front of us; in the meantime, we are witnessing the decline of the old ways of living. This is a transitory phase. But it is not the final stage in the cultural shift, and this change is brought about by technology.

The Internet is changing the way we think, act, perceive and remember. It is damaging our perception skills and the understanding of the environment. Our cognitive capacities are invaded by the enormous data being generated on the Internet, and indeed, they are being narrowed down for use by data miners, who sell it to those who want to influence our behaviour. That is a different issue altogether.

One important area, where digital technology is set to influence our brain, is how we build our network, how we build relationships and understand what it means to have a social relationship and emotions. Social media will have a far-reaching impact on the social interactions, relationships and networking. It affects our understanding skills, memory and emotional skills.[1]

In the beginning, people anticipated that it would bring in some positive changes into our societies, and people would find a space, which they were deprived of by social structures. As time passed by, we also started seeking the same technology as a solution to our own pitfalls and misgivings. Now, many people are trying to replace robots and the Internet with social relationships. For example, the use of sex dolls and toys for companionship.[2] We treat the Internet and other digital technologies as real people. Instead of talking over phone, we send text messages. Instead of communicating with someone in real time, people update their status. This is how we are digitizing their relationships and automating our relations.

Our perception about other people is closely linked to their presence online. People become more real on the Internet than in real life. If you want to understand someone, then a tour of their social media profile is the key to their identity.

Many of us prefer relationships in the digital space to that offline. The increasing reliance on technology is altering what it means to be human. We are already filtering companionship through machines. In the days to come, we may feel obliged to accept robots as companions. Such is our addiction that many people don't know how to deal with people and their emotions other than communicating via technology. Real people are those who engage in activities on others' profiles. Then, the source of a relationship is organized on platforms such as Facebook, WhatsApp, Twitter and YouTube. Teenagers who have been exposed to the Internet since childhood are unaware of the social skills learned by the generations before them. People exchange thousands of text messages every month, spend hours on Facebook and WhatsApp each day. Many of us appear to find life mediated by cyberspace more alluring than real life.

The Influence of the Internet on Teenagers

Social media plays a great role in the lives of India's youth. They are growing up with tablets and smartphones, and they finding solace on Facebook and other social media platforms. They are not even able to remember a time before the advent of the Internet or relate to a relevant social issue. They frequently post about their anxieties and preoccupations, which have mostly to do with their image and acceptance on social media sites. What is ignored is how social media influences their brain development, including memory, responses and processing of information.

The correlation between social media and physical development was well documented by Aaron Bryant in a research paper titled *The Effect of Social Media on the Physical, Social Emotional, and Cognitive Development of Adolescents*. The paper found links between body image, academic achievement, self-esteem and the use of social media. The issue of cyberbullying and its connection to social media was also reported in the paper.[3]

Certainly, the pre-Internet generation had the advantage of fewer distractions; they had real conversations, and their emotions were not hidden behind their screens. However, the privacy of the social media generation is virtually non-existent. Any content posted is easily re-posted and subsequently viewed by third parties, which can sometimes lead to rejection from groups or employers and even abuse.

Teenagers are exposed to images and status that portray unrealistic perfections. Some people post images where they look extremely beautiful, and this is done with the help of technology. Unrealistic and unattainable beauty standards set by social media deflect people from reality. The unrealistic images being uploaded as status and the responses received on them turn us into addicts of social media. Technology exposes their inner conscious of wanting to be noticed, recognized and accepted. It can make people more insecure and constantly vying for support. The overtly sexual representations of their body are in conflict with the real-life situations, which leave them confused between the real and the unreal. The ongoing exposure to unrealistic beauty standards through social media affects how people perceive their own bodies. These perceptions negatively affect their physical and mental health.

Easy and quick access to any number of porn sites affects a person's sexual behaviour. The unrealistic portrayal of sexual acts on professionally produced porn sites negatively affect the minds of teenagers. Exposure to pornography on the Internet, which portrays unrealistic or harmful sexual behaviour and stereotypes, negatively influence teenagers who are just beginning to develop their sexual identities. This may result in them having unrealistic expectations in their sexual lives, which may lead them in forming unhealthy sexual relationships or developing risky sexual behaviours.

Using social media greatly disrupts sleeping patterns by compelling teenagers to stay up late or keep waking up throughout the night to check notifications. The blue light from mobile phones can also disrupt sleep cycles. Many have reported the difficulty in learning and concentrating on studies due to the excessive use of social media. They feel stressed or anxious unnecessarily. They get easily disturbed, annoyed and are becoming short-tempered. It also affects eating habits and people tend to put on weight.

Excessive use of social media also leads to mental health problems such as psychological distress and social media depression—an emotional disturbance associated with the use of social media. Quoting various studies, a report by the Child Mind Institute stated that children are growing up with low self-esteem and more anxiety.[4] The use of social media in adolescents is centred on one common theme—staying connected at all times. For example, LAS Shapiro and G. Margolin found that 73 per cent or more of all adolescents use social media.[5] Teenage girls are particularly susceptible to peer-pressure, and therefore, are at a risk of having negative social media experiences that can affect their development and lead them into anxiety disorders and depression. The Child Mind Study of over half a million eighth through twelfth graders found the number of children with symptoms of depression had increased by 33 per cent between 2010 and 2015. In the same period, the suicide rate of girls in that age group has increased by 65 per cent. It also added that smartphones were introduced in 2007, and by 2015, 92 per cent of teens and young adults owned a smartphone. It also said that the rise in the depressive symptoms mentioned earlier is related to the adoption of smartphones during that period.

The promise of new technologies have already nurtured a culture of social diffusion, intellectual fragmentation and sensory detachment. Technology is enslaving us. To recall our abilities to focus and pay attention in the age of total disconnection and distraction is the necessity of our time.

Endnotes

1. Nicholas Carr, *The Shadows: What the Internet Is Doing to Our Brain*, New York: W W Norton and Company, 2010.
2. Sherry Turkle, *Alone Together: Why We Expect More From Technology and Less From Each Other*, New York: Basic Books, 2010.

3. Aaron Bryant, 'The Effect of Social Media on the Physical, Social Emotional, and Cognitive Development of Adolescents', 2018, *Honors Senior Capstone Projects*. 37. https://scholarworks.merrimack.edu/honors_capstones/37.
4. Rachel Ehmke, 'How Using Social Media Affects Teenagers' https://childmind.org/article/how-using-social-media-affects-teenagers/, accessed on 15 May 2019.
5. LAS Shapiro and G. Margolin, 'Growing Up Wired: Social Networking Sites and Adolescent Psychosocial Development', *Clinical Child and Family Psychology Review*, 2014, 17(1):1–18, DOI: http://doi.org/10.1007/s10567-013-0135-1.

9

Scam Artists

Indians online seem to be in a world full of scams. According to a multi-market survey conducted by the Telenor Group on the impact of scams on Internet users, aged between eighteen and sixty-five in India, Singapore, Malaysia and Thailand,[1] 39 per cent of net users are victims of 'work from home' frauds and 25 per cent have fallen prey to lottery scams. Eighty-five per cent of the country's Internet user base is familiar with the term 'Internet Scam'. Scammer or scam artist is the term used for a criminal who attempts to cheat other people for money. They often have fake identities, operate from a remote place and contact other people over the phone, email, networking sites or fake websites. They build a wide network and succeed in convincing a smaller number of victims, which is more than enough. Those falling prey to this type of predators is the barometer for the other side of the Internet and the mindset of our society and its people.

Con artists, or con men, are criminals, who try to cheat others for money or their possessions, but they do so using their own identity. They are in close proximity with the victim, unlike scammers. The best method they deploy to cheat people is to script a convincing story and instil trust in them. They evoke either sympathy or greed in their victims, with their story and performance.

A scammer or a con artist is not entirely exclusive to each other. There are two different ways to build trust by both of them. A con man builds it in-person, whereas a scammer adopts a fake identity, but both target money and valuable possessions of their victims. It must be admitted that, in general, a scammer is a more modern and appropriate term for a less direct and personal approach to the same idea of a crime committed through deception in the age of the Internet.

Scammers operate in a wide variety of ways. They adopt different methods and strategies. Their actions are well calculated. These are the different types of scams people frequently encounter on the Internet.

1. Clickbait Scams

A post instantly pops up on your social media feed which informs you that something has happened last night or something is going to happen somewhere. It generates curiosity. This is a psychological approach taken by the scammer to manipulate human curiosities. We have a natural inclination to know more, to see things and be at the centre of attention.

Clickbait scammers ask you to click on the post to know more. Most clickbait-type links have catchy or provocative headlines that are difficult for most users to resist and often have little or nothing to do with the actual webpage. For example:

1. 'This young girl went to hug a tiger. What happens next will blow your mind.'

2. 'These facts about slum dwellers will change the way you look at life forever.'
3. 'What this little boy can do with a shovel will make you sob uncontrollably.'

As you click on a photo or weblink, it redirects you to another screen, which instructs you click again on a link to install something. The moment you click the link, the antivirus software in your system informs you that it has blocked a virus. Most people have no idea what to do next. People, who are aware of clickbait scams, immediately log out of the social media account, close their browser and scan the computer within no time using the antivirus software. The scan then finds no threats.

So, one can log back into their social media accounts, change their passwords and reset the security questions. Imagine the consequences of not knowing the intention of such scam artists—your privacy is lost!

So, what exactly is clickbait? It refers to the content used as baits for theft. They are designed to lure people into clicking on certain links, pictures, news headlines, etc. A clickbait does not provide any information about what the link is about; it need not be relevant or factually correct. It just wants to get people interested to click on the link.

Clickbait when executed correctly is one of the best ways to get people's attention. Thereby one can deceive people for money and other things.

2. Impersonation Scams

Let's assume you have received a direct message or an email from someone you know and trust or from a seemingly credible source. The message said that the sender had lost their purse and other documents while on vacation in another country and they needed an $800 wire transfer to pay for their hotel room and airfare. Impersonation scams are those where a person hacks into another's social media account and impersonates to be him or her. The hacker uses it to send fake messages with intentions to deceive people. Its immediate victims are teenagers obsessed with social media, who think the world to be what the social media portrays.

It also works in some other way. Threat-based impersonation scams are common and can be traumatic for the victims. Scammers pretend to be government agencies or reputed companies. They pretend to be from well-known, trusted businesses or organizations, visa application centres, energy or telecommunication providers, immigration departments, banks or law enforcement agencies. They may call and ask for remote access to your computer or other relevant usernames and passwords to fix a problem. They may email you fake invoices or fines and threaten to cancel your service or charge excessive penalty fees if you don't pay them immediately. They may threaten you to pay a fine to avoid disconnection, legal action, arrest or deportation.

3. Dating Scams

We all receive emails or messages from strangers and often don't verify their authenticity. This is especially when you believe things on social media the way they are presented. Let us assume a scenario where you have received a message or accepted a friend request, and the stranger tells you that he is an Italian living in France. After exchanging messages, both of you plan to meet. The stranger then sends money to your bank account for the cost of air travel, but then, he also asks you to send some money back because they have been laid off from their job and need the money for rent or some other unforeseen expense. You have deposited the cheque and wired the money. However, soon you are contacted by his bank that the amount deposited was by a counterfeit cheque and you are asked by the bank to repay them the money you transferred to the stranger. On top of losing the money, the stranger has disappeared and you have never heard from them again. This is called a dating scam.

4. Identity Theft

You have seen a post on social media feed that says, 'Win a two-wheeler'. Without a second thought, you click on the link and are directed to a website that asks you to answer some questions. It also asked you to enter your name, contact address, telephone number, bank account details, ID proofs and their scanned copies to enter the contest. How many people would realize that this information could be used to steal your identity and there is something called identity theft? How many people close the browser and log out of their accounts? Many teenagers are unaware of identity theft and privacy violations.

Often, you have been called from the banks alerting you to the fact that your balance is zero because you purchased something online! Shocked? You, who hardly uses your debit card, realizes that your financial records have been stolen and misused. Identity theft is the most common of cybercrimes, with people illegally obtaining personal information such as bank account number, Internet banking passwords, Aadhaar number, credit card number, etc. The thief can then use the stolen information to pose as the victim for financial benefit or to commit a crime.

Stealing financial data is one form of identity theft. Cyber criminals can steal any of your identity markers from your general data such as address, email or phone number to more specific details like tax returns, credit card information, bank account details, photos and social media accounts.

Identity thefts can have severe consequences. For instance, theft of one's Aadhaar card can be seriously damaging as it can be misused in various ways. Most identity thefts are aimed at stealing your financial data or email passwords. Identity can be stolen in several ways, right from injecting a malware into your system to someone physically looking into your unattended device.

Identity theft can harm you in various ways. It can lead to financial loss, in case of someone stealing your bank data or even public embarrassment in case of someone hijacking your social media account or circulating your private photos on the Internet.

5. Phishing Scams

You receive an email that says it is from your social media website, whose services you have been using. The email informs that your social media account had been locked for some technical reason and asks you to 'verify' your account by clicking a link provided by them. After you have clicked on the link, you are redirected to a webpage that looks similar to the social media website you have been using for sometime now. The webpage instructs you to enter your username and password to verify and unlock your personal social media account. What happens when you have entered this information before realizing that you are on the wrong website is that your passwords, username and access to your activities on your social media are sent to an unknown source. Your account can be misused in a number of ways. Once you lose this information to someone, you cannot recover your profile. These are called phishing scams, which are least known to ordinary users of the digital space.

6. Sweepstakes and Lottery Scams

One day, a message pops up on your personal social media profile. It could also be a short message on your mobile. It says that you have won a huge sum of money in a lottery. Then you are told that you need to send some money to pay the taxes and fees to get the prize. Many remain unaware that lottery scams work by asking you to pay some sort of fee in order to claim your prize or winnings from a competition or lottery you have never entered.

Frequently, people receive notifications that they have won a lot of money or an amazing prize in a competition, lottery or sweepstake that they don't remember even entering in. The message usually comes via email, mobile phone, text message or as direct messages on social media platforms.

The prize they said you have won could be anything—from a holiday to a beautiful overseas location to electronic equipment such as a laptop or a smartphone or even money from an international lottery.

To claim your prize, it is compulsory, according to the organizers, that you pay a fee. Scammers will often say these fees are insurance costs, government taxes, bank fees or courier charges. The scammers make money by continually collecting these fees from you and stalling the payment of your winnings.

The email, letter or text message you receive will ask you to respond quickly or risk missing the opportunity. It may also urge you to keep your prize private or confidential, to 'maintain security' or stop other people from getting your prize by mistake. Scammers do this to prevent you from seeking further information or advice from other sources.

Lottery scams may use the names of legitimate overseas lotteries, so that if you do a superficial research, it would not seem to be a scam.

How many people really understand these nuances of digital media, especially those who haven't entered into any lotteries or online contest? Most ordinary users are unaware of this danger and fall prey to the dishonest intentions of someone unknown.

7. Work at Home Scams

You receive a message from someone who has claimed you could make thousands of dollars from home every day if you invest a small amount of money in a programme. After you have expressed interest in the programme, the person asks you to share your credit card number to pay for a start-up kit. You then share the required information, but you never receive the start-up kit. When you check your credit card account statement later, you discover that, in addition to the fee demanded, some unknown sources have charged you a monthly membership fee.

Jobs are difficult to get in our part of the world. You may have scored well, but the market is such that there aren't many jobs for you. You panic, and there comes a time when anything works for you. Here comes the sociology of the 'work from home' scam. It is one of the most popular keywords on the Internet. Many people are looking for 'work from home' opportunities because of the inconsistent job conditions in the market. Many companies nowadays promise an opportunity to work from home but are often fraudulent.

Since we are a largely patriarchal society, women prefer to work from home or are encouraged to take it up. Hence, are likely to fall prey to dubious work from home offers.

Many data entry and transcription companies are dubious. They have impressive websites and hold seminars in five star hotels but demand a down payment for people to register with them and promise a fixed income for every month month, which sounds tempting. Once you register, these firms might even pay you for a month or two before disappearing completely.

Suppose you sign up for a transcription firm online. Your work seems easy and you do get a couple of calls. By now, you have shared your bank account details to remit the payments. However, there is no response from them and you find out that the website has shut down. Your emails go unanswered, and of course, you never are paid. This sort of cheating is prevalent in the cyberspace.

8. Spoofing

The word spoof means to hoax, trick or deception. In the Internet world, spoofing refers to tricking or deceiving computer systems or other computer users. This is typically done by hiding one's identity or faking the identity of another user on the Internet. Spoofing takes place on the Internet in various ways. The most frequent and common method is via email. Users send emails to the targeted user from a bogus e-mail address or fake the e-mail address of another user. For example, you may receive an email from the address of a person you are familiar with.

Spammers send messages from their own Simple Mail Transfer Protocol (SMT), which allows them to use fake e-mail addresses. Therefore, you may receive e-mails from an address that is not the actual address of the person sending the message.

Another way of spoofing on Internet is Internet Protocol (IP) spoofing. It involves masking the IP address of a certain computer system. IP spoofing makes it difficult to track the source of a transmission. This may cause the server to either crash or become unresponsive to legitimate requests.

Then, spoofing is done by faking an identity, such as an online username. For example, while participating on a web-based discussion forum, a user may pretend as the representative of a certain company or group. However, they are, in reality, no way associated with the firm or group. In online chat rooms and social networking sites, users fake their age, gender and location.

While the Internet is a great place to communicate and connect, the fact is that there are users who hide their identity and engage in crimes, acts that can have devastating consequences.

9. Cybercrimes

Many people have fallen prey to online fraud using popular e-commerce sites including Amazon. A spoof site exhibited a 99 per cent discount offer for products on sale. A click on the link takes users to a website that almost looks like the Amazon webpage at a glance. We trust whatever is seen in cyberspace as real and don't give credibility a second thought. People never take time to see that it was a blogspot. No official Amazon page displays any such offers.

Spoofing the online identity of a famous business brand or organization to deceive people is now common. It fakes the identity of a real organization and deceives people for their money or jobs. Even human trafficking is being done through spoofing.

Some users leak usernames and passwords of bank accounts in order to track credit and debit card details of bank account holders. Even spoof sites are infiltrating into Google Play. In fact, fake apps have reportedly been launched that claim to be leading banks in India.

Since India's digital and Internet literacy is poor, the country is a hotspot for scam artists. As people simply share and forward almost all the information being shared on the social media platforms, the risk involved is very high.

Scam artists on the Internet are spoiling the nature of cyberspace, making the youth and others fall prey to their dishonest intentions. Scammers on the Internet source your identity, money, user activities, privacy and access to social media account, which, in turn, are used for endless purposes, including anti-national activities.

10. Romance Scam

The online dating industry is now a billion-dollar global industry. Romantic relationships are fast catching the attention of apps and the like. Many youngsters in India are active members of dating sites. As the number of people connected to the dating apps on the Internet are increasing, the opportunity to meet new people online has been more compelling. The nature of dating has changed exponentially over the past decades. While earlier, one would not admit to being on a dating site, today it is much less contemptuous.

Although it is common to find people forming relationships over the Internet, the problem it has created is also quite complicated.

Frauds have existed since ancient times. We have heard of stories of deceit and treachery involving men and women who pose as wealthy individuals promising love, and even marriage, to gullible victims. Once they have exploited the victim financially or sexually, their disappearance has found place in love stories. However, why it is even more important to talk about it now is due to the alarming number of such cases reported. A number of homegrown dating websites and mobile applications have taken the market by storm following the instant success of international dating app Tinder in India. In order to capitalize on the changing lifestyles and love choices of millennials in the country, many international competitors too have launched themselves in the Indian market. However, the real beneficiaries are fraudsters who prey on innocent victims seeking love and companionship on these platforms.

Users trying to dupe someone by getting involved with them romantically signify a pattern that typically corresponds to our peculiar social system unlike the romance scam in other parts of the world. The victims are usually middle-aged men and women from upper-and middle-class families who are well educated and have access to Internet. Divorcees and those experiencing some kind of emptiness and loneliness turn to these websites in search of a partner but fall prey to these scams.

Once the scammer contacts the target on one such online platform, the acquaintance is then taken forward on other social media platforms such as WhatsApp and Facebook, where fake accounts have been created to match their dating profiles. Once the scammer has gained the trust of the target, money is demanded under the pretext of paying medical bills of a sick parent or on promises of marriage or future dates. Soon after, the transactions are made, and the impostor disappears and becomes untraceable. This is the usual story nowadays.

Alchemy of Internet Scammers

Why India is prone to Internet scams is also an answer to the vulnerabilities of our time. Scams tell us the psychosis that eats into our society. Greed incentivizes our behaviour. Fame induces this behaviour in people and people crave for attention. They prefer to be at the centre of events. Scams appeal to our weaknesses and sensibilities. It tells how much greed and luxury we are still obsessed with. Scams tell us a lot about the nature of our society. It is an index of the social psyche and the irony of our time. More than the gullibility, greed or ignorance of the people, scams are the barometer of relationships in our society—our weaknesses, rabbit holes in the social fabric and how easily people can be misled.

People try to do anything for love, money or fame. Short cuts are welcome and this has been normalized in our e-society. Scams and our social structure have close interconnection. Many scams on the Internet are closely linked to and even reflect the gaps in our social system. These include caste system, religion, family and relationships. Our narrow approach towards love and relationships means that it is easier to fall prey to dating scams.

Despite all the media coverage and public awareness, there is only one answer to the question why people are still falling for scams. It is the human nature! Scammers know it better than the rest of us. This is capitalized by scammers to emotionally lure children and teenagers into their traps. There are people who are longing to amass as much money as possible, which they don't even need. Scammers know it and they cleverly manipulate this insane human attitude of wanting more money. If your perception of beauty is still ruled by socially accepted norms of beauty, chances are that you live in a stigmatized society. This, too, is well known to scammers; hence they are more likely to manipulate you. There is a clear psychology behind Internet scams. They target our emotions, but they do not question our intellect, knowing that we are more tied to our culture with our emotions. Scammers look for specific groups, including children, the elderly, lonely people, divorcees and women.

They exploit the lack of awareness and social isolation. Scams are successful precisely because the predator understands the weakness of the prey. But targeting what ails the society and leveraging on it, scammers emotionally manipulate victims. The police and legit companies warn the public that their representatives neither ask for payment through a telephone call nor make unsolicited calls asking for personal information. But many psychologists are of the opinion that the victims may be too afraid to think rationally and respond to such information.

Scam victims may be distracted or looking for shortcuts. Social media has already adversely affected people's eye for detail, contemplation, evaluation and other cognitive abilities. Social media has indeed created a generation that is obsessed with self-gratification and too much attention. The use of mental shortcuts is another reason why scams succeed. People have certain psychological traits which allow them to negotiate daily life more easily without being overloaded with information and decision-making. Scammers exploit these psychological traits to get their way.

For example, if you see a website with the logo of a government department, you often do not question its authenticity, and scammers have been able to get away with passwords and other sensitive information through these fake websites. Other shortcuts people rely on include social compliance and conformity.

Indeed, scammers use psychological tactics which manipulate the vulnerabilities of people. Their tactics have close parallels with the social structure—the unique family system, caste, religion, class disparity and gender.

10

Love and Betrayal

One of my students, Saida, who graduated in Political Science from the institution where I work as a teacher, once related to me how she had become a victim of betrayal online. A close friend's boyfriend impersonated Saida on Facebook and sent friend requests to strangers and sexted with them too. Finally, he forwarded all those fake chats to the Facebook profile of Saida's boyfriend and shared screenshots on various social media platforms. Her boyfriend blindly believed the content. Ironically, a fake person on the Internet was more credible than the trust they had built over three years. Their broke up and Saida was terribly upset. She never got back into that relationship even though finally she told her story. Saida was a victim of 'revenge porn' on the Internet and this is not an isolated case.

Many such cases are being reported, but they constitute a tiny fraction of the actual number of incidents. One has heard of incidents where men, educated and employed, in reputed firms or friends, post nudes on the Internet without the knowledge of the women concerned. Such incidents happen frequently, but are mostly unreported because of the shame involved. For those reported, most cases carry a common thread: an educated boy from a respected family posts either nude pictures or videos of his ex-girlfriend on the Internet from fake accounts. In other cases, boys morph the girls' face and make the picture go viral in order to tarnish the girls' reputation and make their lives miserable.

What is scarier about such crimes is that this is happening even as you read this book. The said content is available on plenty of platforms and can still be accessed by anyone. This is why it is traumatic for the victims.

The nature of such acts exposes the brute mindset of people and shows us where the Internet is headed. It is also a measure of the destructive personality of the Internet. What is disheartening is that on the Internet, a brute mentality is shamelessly vilifying women. It is indeed shocking to know that people seeing the content are known to victims, including colleagues, bosses, subordinates, friends, parents, children.

Revenge porn involves a brutal act of sharing explicit material, including photos and other obscene material on the Internet without consent. The motivation stems from wanting to seek revenge by causing emotional distress to current or former partners for some perceived insults. Sometimes, the material being posted on the Internet involves those produced by way of hacking or spy cam. Recording equipment are placed discreetly in toilets, wardrobes and similar places, where the subject never knows they are being filmed until it has been posted in cyberspace. The crime has also been referred to as non-consensual pornography, a nasty turn in the biography of the Internet. Other instances are when people have recordings of their conversations when they were in a relationship. These recordings may have been done with a vicious intent to use it as a tool to blackmail the other person. Once a relationship gets strained, the recordings are sent to the Internet for public shaming. Revenge porn is a misleading term because in it, private moments or images are captured without the prior knowledge of the subject. In pornography, one may assume that the consent of the actor has been taken. The term had gained popularity in recent years as the motive of the offender is more often aimed at taking revenge on the victim for breaking up or rejecting a relationship or shame resulting out of a perceived insult. The offender then uses intimate or morphed images to shame the victim.

The point is that this sort of content is posted on the Internet as it is a place where any content can be shared from anywhere anytime. They can also be posted on forums and other less regulated platforms on the Internet. But it must also be taken into account that app developers such as Facebook and Twitter are clamping down on the practice. Pornographic content is banned on social media platforms such as Facebook, Twitter, Instagram, Reddit, Tumblr, Yahoo, Google, Blogger and Bing. Once reported, the site will delete non-consensual pornographic images.

Revenge porn is one of the grave forms of cybercrimes. Cybercrimes, in general, take full advantage of the personality of the Internet such as anonymity, speed, secrecy, reach and interconnectedness. The impact of this type of crime is attaining alarming proportions. Its effects touch just about everyone to some degree. Revenge porn on the Internet shows a portrait of our culture. Men use social media platforms with vicious intentions to bring disrepute to women in a society, which doesn't treat them as equals. Men want to be revenged when a relationship ends, due to marital discord and feelings of inferiority. Revenge porn is often detrimental to women as the 'exposure' is often hard to survive mentally. While men involved in many relationships are seen as normal and masculine, for women, this is immoral. Their character is assassinated. Their reputation is damaged. Their body is slut-shamed. Their intimate life is exposed. Most obviously, the shares, likes, comments and replies furnished to such spiteful and ferocious content bark many unwanted consequences.

Young people are the prime victims of revenge porn. The culture of retribution has almost become routine among the youth. They do it for silly reasons, including personal conflicts, jealousy, professional rivalry and frustration out of being let down in relationships. Video hosting services have been more disparaging. They are used to commoditize and sexualize women's bodies. Video hosting platforms have provided space for uploading videos of intimate moments. In some cases, men involved in such heinous crimes can't be identified at all.

The nature of this type of crime itself is sexist. So far, there are hardly any cases reported in which men are victimized by revenge porn. In most cases, the atrocious revenge culture is deployed against women.

The cases being reported signify the fact that a new culture of revenge porn has emerged in India, which has serious consequences on our relationships.

Although a peculiar form of crime culture has developed on the Internet, revenge porn is a cultural crime too. It has never occurred in the initial years of the Internet and has occurred very recently.

Although such events could be avoided or prevented, and if found guilty, perpetrators could be penalized. They have been occurring repeatedly, and in most cases, they go unreported. The accused are booked under IPC Sections 509 (word, gesture or act intended to insult the modesty of a woman), 419 (punishment for cheating by personation), 469 (forgery for purpose of harming reputation), 500 (defamation), among others, and relevant sections of the IT Act. There are very few legal provisions pertaining to crimes on the Internet. However, no laws specifically mention revenge culture. Especially the IT Act 2000, which is arguably the sole legal measure that addresses such crimes, does not mention it.

Love and Betrayal

Most Internet-based love affairs tend to be short-lived, but the trauma they carry are not. Internet provides ample opportunities for many people to date online, but there are some people who use Internet as a tool to dupe, emotionally or financially, those who are ignorant and vulnerable.

The Internet is used as a tool to bully, harass, stalk or intimidate a lover and bring financial loss as well as emotional distress to them. It is easy to betray people by the way of fake love affairs. Some people easily fall prey to the trap of trained offenders who know how the Internet works to their advantage.

There are few things that tend to make cyberspace a hostile place. The fact is that impersonation is such a common thing on the Internet that criminals deploy as a moneymaking tool and advance sexual favours from victims. It is disturbing to find out you have been duped by someone you trust is even worse. When a girl realizes she was befooled, she is unlikely to report it fearing the embarassment it might lead to.

When it comes to a romantic or intimate betrayal, it undermines the unspoken promise of such relationships. No matter what, the person you love and trust will care about your well-being and never hurt you intentionally. Indignation is one of the most common emotions experienced by the betrayed party.

For this reason, cheating has caused distress of various sorts.

But one thing is clear that those people are catfished, which is a much more modern phenomenon. It is comparable to the betrayal or infidelity. It might be worse. Victims of Internet romance fraud are not only betrayed, they are insulted and disrespected to a degree far greater than cheating.

For as long as the Internet has given people the freedom to communicate with one another behind a wall of cyber anonymity, there have been reports of many forms of violence. Although some romantic affairs on the Internet may be genuine, many just lead to awkward philandering. Love on the Internet can seem easier for some people than going for a traditional love affair due to reasons including safety and distance. You would not want to feel pressured or vulnerable, and the Internet helps take care of those things.

However, online relationship can often lead to financial damage and heartbreak. There are a series of reports that women are increasingly victimized. Both men and women have their own justifications to get involved in relationships, even socially non-conforming ones, that too with anonymous people. In a world where fraud, deceit and cheat have become norms, certainly love too has myriad forms.

Men and women befriend hundreds of strangers on social media. With hundreds and even thousands of strangers as friends, online romancing has created consequences that are often difficult to interpret.

There are also love scam artists, otherwise called 'catfishing', who use cyberspace to lure people into a relationship, take money from people and put them through emotional trauma. People who use cyber romance as a hoax typically create fake Internet profiles designed to lure people into love affairs. They may use fictional names, claim to be celebrities, film artists, musicians or be known to famous people, or falsely take on the identities of real, trusted people such as industrialists, military personnel, human rights activists or professionals working abroad.

They express strong emotions for you in a relatively short period. They try to build trust in you. They build up stories after stories to gain your trust by way of showering you with loving words. They share personal information and even try sending you gifts. They may take a few months to build what may feel like the romance of a lifetime and may even offer to travel to visit you.

As the trust grows, they request to move the relationship away from the cyberspace to a more private platform, such as phone, WhatsApp, Skype, email, etc. Once the trust starts to build in the relationship, they may raise it to another level to make it more intimate. They may also ask you to send pictures or videos of yourself, possibly of an intimate nature.

As you advance into the relationship and want to meet the person, they will take the relationship to yet another level. There is a time when you cannot reject the lover. Now that you trust the relationship and you feel that there is no more 'distance' between the two, they will contact you for money, gifts or your bank details.

Internet romance poses a risk to personal safety, as there are people with a criminal mindset. Passwords of Facebook, Instagram, email and other vital information relating to digital identities are being exposed to criminals and that puts their victims in dangerous situations that can have tragic consequences.

It is doubtless that the way we express our love to others, and our relationships, have changed due to the Internet. There is a social system, which makes people more likely to look for solace online, but unfortunately, most people are unaware of how the Internet can be misused. There are reports of apps being developed, including One Love MyPlan, that spread awareness among users about safety plans in case of an abusive or threatening relationship. Yet, the question is, how women-friendly are the cybercrime laws? Do they address the peculiar gendered nature of Internet culture, and issues such as revenge porn, love, betrayal and misogyny on the Internet? The Indian Penal Code, The Indian Evidence Act and The IT Act 2000 are usually evoked to deal with acts of revenge porn, misogyny, love and betrayal. Yet, one fears that even strict laws are not sufficient to check the ills of this patriarchal society.

11

Censorship and the Changing Personality of the Internet

It is a truism that the personality of the Internet has changed. It is torn apart by things that have already divided society. Professed neutrality is a subjective idea here. The Internet has not become a neutral social space as one would have hoped. The social architecture that invented it has had an oscillating power over its personality.

Evangelists of the Internet had initially foreseen that it facilitates contact between like-minded people, brings social change and even repressive political systems were supposed to be stonewalled. The expectations about the Internet were such that people began to romanticize its potentialities and even see it as god.

Countless evidence, however, suggest that the stories of the Internet are neither that hopeful nor straight. People have, in due course, developed a mixed reaction. Now it seems that it is being divided by many things. It has become a playground for criminals.

Social structures influence the shallow personality of the

Internet, which is more vulnerable to divisive forces now. The factors that are linked with identity, such as gender, class, race, region, caste, religion and shape, rile the real character and operational dynamics of the Internet. Its real fabric being subjective to its makers, the Internet cannot sweep away the bigotry of the social ecology in which it is contemplated.

When talking about the personality of the Internet, it is useful to have two different perspectives of what the Internet is—that of the user's and the creator's. The first idea is to view the Internet as a technological invention. This constitutes the idea that you upload data at one end and it is delivered at the other. The second is the perspective you and I can relate to.

However, no matter what, one thing that is unique to the Internet is that it doesn't understand what you are doing and cannot be distinguished as good and bad, a speech that is hateful from a speech that is about freedom.

The Internet is just a medium. It carries your data from one place to another. And the data remains there as long as you wish it to be. Until and unless the data carried out there is not removed by the service providers, it remains retrievable anytime anywhere. What gives it a personality is what you do with it.

Since, the Internet itself cannot distinguish between a bad behaviour and a good behaviour; the consequences of any digital platform are that it cannot prevent the bad behaviour. It cannot censor you nor promote you. It rests on the human agency to handle it.

This peculiar nature of the Internet indeed has some implications. You voluntarily access the Internet. There are service providers, who provide you with facilities to access the Internet. For example, BSNL, MTNL, Reliance Communication and Idea are big players in this aspect. These players are not the Internet per se, but provide the facility to reach Facebook, Twitter,

Amazon, Uber, Google and others. They are specific platforms, which, from our own perspectives, are what the Internet is. Those providing services would consider the Internet a free, liberal and open space. For most of us, the Internet is about the applications we use. But the Internet service providers are not in the application business. Although they would love to be in the lucrative application business, they are in the business of data transfer. So, all the problems that arise on the Internet are not directly related to the service providers, such as BSNL or MTNL. The problems we confront on the Internet take place when there are applications, or platforms, including Google and Facebook.

Indeed, the personality of the Internet is what people do on these platforms and how they do it. Hence, what constitutes the personality of the Internet is more than a scientific question. The Internet is deeply embedded with our social system. It intimately connects with the culture and social practices. The social architecture deeply influences how people behave on the Internet. And the Internet, in no way, controls or influences that behaviour, since it is just a medium.

Do We Need To Regulate the Internet?

The world behind the screen where all of us stay connected with one another also distances us from reality. We assume the screen to act as a cover against the digital world we are exposed to through our desktops or smartphones. This makes us think that we have the license to do anything out there. When people are behind the screen at their homes, they tend to lose all their cherished moments that they shared before the Internet entered into our lives. Since people are not emotionally connected with everyone on the net, they have the courage to say anything to anyone. Civility is lost; sociability is robbed; understanding is ruined.

I can cite many instances in which people pretend to be gentle, secular, loving, helping and have so many other enviable qualities. On social media, they present themselves as the best people. On the Internet, they are the Good Samaritan, helping others and have all other repositories of great virtues. But as you keep looking deeper, the same people can be quite the opposite. You can also spot them misusing public office for their personal gain and promise to end corruption. I know people who purchase a laptop or a desktop for twenty five thousand from the public fund but get it billed at forty thousand.

What I believe is that we live in a society that is already hypocritical. We live with a double morality; one that is showed off and the other that we hide. It is in this second morality that is buried in the dark; technology has made us narcissistic and less engaging.

The Internet has become an embarrassing factor in our social lives. People bully their perceived enemies behind the screen, tarnish their reputation, spread rumours, gossip, etc., which they would not do in public. This is because of the second morality; we fear the first morality that is open to the society. It is constructed by the society and we try our level best to conform to it. But we are persuaded by our second morality that is hidden inside us to do those things which our actual nature wants us to do. The Internet is just helping the second morality hidden inside us. We are constantly tossed between the two.

As the Internet cannot remain neutral to the social system in which it is invented and put to use, it too reflects its social alchemy. What people habitually do in real life is reinvented and carried over to the Internet. Indeed, the Internet is what the people who make use of it are. It has information centric networks. This is where it becomes important to discuss the significance of Internet regulation.

Data Censorship: A Comparative Perspective

Governments, both in democracies and authoritarians, increasingly vie for data censorship on the Internet. The Google Transparency Report[2] says that Google receives zillion of requests from governments across the world to remove objectionable contents from its products and services. Internet censoring has attained equal uneasiness across western democracies recently, though it was initially an issue raised in political regimes other than liberal democracies. Electronic Frontiers Australia[3] prepared a report titled *Internet Censorship: Law & Policy Around the World*, which deals with Internet censorship in various countries. The report says that there were attempts since 2005 to curb the Internet and its contents owing to various reasons in different countries. The extensive report is a testimony to the fact that there is huge vagueness about the architecture of data censorship regime across political systems. Indeed, vague attempts at censorship and regulatory frameworks leave lovers of freedom on the Internet confused.

Political censorship by governments, such as China, Iran, and countries embroiled in the Arab Spring have been inspirations for other nations' governments. Syria, for instance, banned iPhones.[4] YouTube, WordPress and Blogger have been banned in Turkey, along with 138 search words on the Internet.[5] Iran issued regulations to crack down on Internet expressions.[6] Belarus has outlawed browsing foreign websites.[7] Denmark outlawed anonymous access to the Internet.[8] In the US, Stop Online Piracy Act (SOPA) by the House of Representatives and the Protect Intellectual Property Act (PIPA) by the Senate, according to critics, would kill freedom of expression and allow the government to censor the Web.

Reporters Without Borders identified a list of 'Enemies of the Internet' and a list of 'Countries under Surveillance'.[9] To get an understanding of the world map of data censorship, one may refer to the Boy Genius Report.[10] The map uses the criteria to distinguish the level of censorship, and issues are split up into four main categories: 'human rights violations,' 'freedom on the Net', 'obstacles to accesses' and 'limits to content'. It shows that Iran, Syria, Cuba and Egypt round off the top five countries with the most Internet censorship in the world.

Governments across the world, no matter how democratic or authoritarian, want to censor the Internet. When the access is regulated, it is often not done in the real spirit. The question remains—whether those who misuse the Internet or the forces such as the government, who regulate the Internet in the view of protecting social decency, morality and culture, are the real enemies of the Internet. It is an interesting question to ponder over, especially for those who are all for the Internet.

The fact is that different countries interpret data freedom differently and service providers interpret Internet freedom differently according to their conviction and national laws in the home country. While different interpretation confuses the attempt to censorship, the right to expression, along with a couple of valuable rights under various treaties, conventions and constitutions, are violated. Although rights are violated due to Internet censorship on one hand, a wide range of Internet behaviour strongly calls for and justifies the attempt to Internet censorship. Other factors, including this conflict, are considered for making decisions regarding Internet censorship across governments of the world.

Regulation Attempts in India

In India, there seems to be a mindset across the political spectre for data censorship; plans are also underway to bring social media sites under the radar. Recent events such as individual activities leading to sedition charges, hate campaigns via social media platforms, events related to Muzaffarnagar communal riot in U.P., online campaigns against northeast Indians residing in south India, send a chilling signal into the government circles to target the Internet.

The Election Commission (EC) on 25 October 2013 issued instructions to the chief electoral officers in states and Union Territories and presidents and general secretaries of political parties regarding the use of social media in the electoral environment. The EC has broadly classified social media into five categories. While doing so, the commission mandated the political parties to get a 'pre-certification' for political advertisements on the Internet, which makes the political class more cautious while using it. The instruction requires political parties to furnish the expenditure for creating social media accounts, salaries paid to the staff that maintain and operate it and the cost incurred to Internet companies. These fall under the election expenses of a candidate. It was a bold decision by the commission and a fair one too.

In 2012, there have been reports about restricting data censorship as the Government attempted to ban selected Internet sites. There were reports that social media posts led to communal tension to target people from northeast India residing in Hyderabad, Bangalore, Pune and Mumbai.[11] Following these incidents, there were discussions in the Parliament in August 2012 on how to censor the Internet. Consequently, the government blocked about 250 websites and social networking sites for spreading inflammatory content that incited panic among thousands of workers and students from the country's eight north-eastern states.[12] When the Centre leaked the information for Internet and Society, a non-profit organization, it came to light that these banned sites also included many popular media outlets.[13] Meeting of the National Integration Council, attended by the chief ministers of various states in India on 23 September 2013 had the opinion that social media is spreading hate; therefore, it must be regulated.

There has been a spike of 90 per cent in the number of demands from the Government of India and courts in the latter half of 2012 as compared with the first half of the year, for the removal of contents from the Internet, says a transparency report by Google Inc. The Indian government, for example, was among twenty countries to request for the removal of the controversial film *Innocence of Muslims* from YouTube, a film that mocked Prophet Muhammad and sparked protests in the Islamic world. From July 2013 to December 2013, Facebook says it has removed 4,765 pieces of content on request from India, the largest in the period.[14] The Freedom House, a US-based independent monitoring group, bestowed India a score of 47 on a scale of 100 in May 2012–April 2013[15] for the biggest drop among all the countries assessed in terms of Internet freedom.

While the government grappled with insufficiency in tackling the censorship of content, such inexperience is seemingly worrying in respect to legal proceedings. Rulings on Internet-related subjects are adjudged as they have been conventionally done, but Internet-related matters need more technical knowledge. This lack of experience can make rulings problematic. According to a 2013 report by *Medianama*,[16] a Department of Telecommunications ordered Internet Service Providers to block a list of 78 URLs, of which 73 were related to the Indian Institute of Planning and Management (IIPM).[17] These included reports by leading magazines on IIPM and a UGC notice pointing towards the unrecognized status of IIPM. Most of these links were critical to the institute.

Political censorship has increased since 2008; an amendment to the Information Technology Act 2000[18] granted the government power to block any content in the interests of friendly relations with foreign states, defence, sovereignty, public order and national security.[19] However, there were no filtering of content in India in 2007 and both types of Internet censoring have been on the rise since 2012, says the OpenNet Initiative.[20] The two types refer to political censorship done in order to maintain friendly relations with other countries. It can be motivated by vested interests of those in power to bring the Opposition, or anyone who dissents, under control and filtering to protect society from their perceived sense of social unrest or decay, including banning porn sites or communal content.

The OpenNet Initiatives reported that Internet platforms such as proxies, websites with information on human rights and content related to free expression have apparently been selected for filtering.

In India, for instance, the Indian Computer Emergency Response Team (CERT-IN) was set up by the Department of Information Technology (DOT) under the amended IT Act to implement a filtering regime.[21] In 2004, CERT-IN became operational as the national nodal agency to review complaints and issue blocking instructions to the DOT.[22]

Following the 11 July 2006 Mumbai train bombings, because the attackers reportedly communicated by means of the blogosphere, CERT-IN blocked websites. These included extremist sites, for example dalitstan.org[23], that reportedly supported the formation of a Dalit homeland within India. It is argued that even though CERT-IN had issued orders to block specific websites, no communication had been made to them in advance.[24]

In recent years, the government has been repeatedly

targeting religious, political and extremist commentaries on online platforms. In addition to the examples already considered, hundreds of more web pages are ostensibly under the government radar due to its communal or religious nature.

Paid news phenomena, as was a tendency more prevalent in traditional media outlets, are now frequently reappearing on digital media platforms. Alleged fake social media campaigns are examples of such phenomena; the phenomena of fake social media advertisements have increased on digital platforms over the past years.[25]

An organized attack awaits those who express opinion counter to the official position. Radical forces are using cyberspace to attack their perceived enemies. Censoring the Internet, though not in the spirit of democracy, is a welcome sign if one considers the fact that radical forces have colonized the Internet space. This is in the view of e-reports that artists, journalists and activists were threatened from speaking or exhibiting works deemed insulting to Hinduism.[26] Certainly, censorship on the Internet is a dilemma on the freedom of data in India.

At Gunpoint

The issue is when free speech becomes hate speech and falls into legal difficulty. Social networking sites such as Facebook and Twitter are open and public forums where users can shoot and upload videos, freely express their views on politics, race, religion and sexuality. They can also create profile pages and groups to join for or against a cause. Content regulation standards are often vague and vary. It is because such practices are usually outsourced to low-paid employees of third-party companies. Facebook and Twitter have different approaches to regulate content. Twitter, self-described as 'the free speech wing of the free speech party', has largely defied any limitations on content by either governments or citizen groups.[27]

The notion of the Internet as a regulation-free medium is very charming in theory. However, experiments in many countries show that the Internet as a 'free space' is confined only to paper and theory.[28] The reality is that the Internet now seems regulated to the scope that each nation considers what is doable and apt according to the national context. In spite of many new communicative and technical options of the Internet, countries such as the US[29] and Germany attempt to fit new media into their aging free speech standards.

It is broadly accepted that hate propaganda troubles society as a whole, and many countries outlaw hate speech in their criminal codes such as Canada, Germany, France, Austria, the Netherlands and Italy.[30] The most popular hate themes on the Internet include white superiority, intimidation of people of colour, Jews, neo-Nazism and Holocaust denial. The most controversial topic belongs to Holocaust denial or revisionism. In democratic societies, any restriction on speech must be in conformity with recognized standards on the limits of freedom of expression. The Internet has provided unique resources for hate propaganda, which calls for serious political resolution as well as regulation. Its influence goes beyond text and word that were available even before the existence leaflets and brochures.[31] Given this historical context, hate speech has primarily been understood in India as referring to speech intended to promote hatred or violence among India's religious communities.[32] In reference to Internet in India, a glance at the user-generated hate content gives a similar impression that is prevailing in Western societies regarding the question of radicalism.

Many countries are on the way to make some regulations in the use of social media; the European Union and Poland[33] have already put some guidelines. For instance, on the backdrop of an anti-Islamic video that provoked violence across the globe, Google considered it was not a hate speech, despite it being asked by the White House to reconsider the decision. Google is not the only one to engage with problems involving anti-Islamic video and radicalism that appeared on YouTube. Facebook blocked links to a video in Pakistan, where it defies the country's blasphemy law.[34]

Measures to check hate on the Internet:

1. There is a need to review hate speech legislation. The law should be a living, dynamic document that leaves room for refinement and modification over time. The law should also have specific definitions in place for social media, such as regulating abusive and threatening language, or if that language was used to stir hatred against a specific group of people. This approach will have its limits because of the problems with its implementation. Neither law nor government censorship may always be a panacea to hate. In fact, it is very hard to create a legal prohibition or prescription against the free flow of information on social media.

2. Social media should have user norms regarding language and decency. This is in view of the unchecked and unscreened social media comments. At present, on this issue of comment moderation, the responsibility is placed with the media outlets and the Internet companies such as Facebook and Twitter.

3. Security experts claim that almost all our activities are under the surveillance radar of intelligence machineries such as Network Traffic Analysis (NETRA) and The National Investigation Agency and the National

Counter Terrorism Centre or NATGRID. On one hand, surveillance comes at the cost of our privacy; on the other hand, privacy and anonymity also create the license to abridge the rights of other people. So, there should be clear demarcation between privacy and secrecy in cyberspace. Secrecy is being misused by criminals, and hence, there should be a mechanism to establish a user's identity before signing up for social media. If he or she is not ready to disclose it, there must be a mechanism to provide the same under certain conditions.

It is a fact that the nature of radicalism in all epochs in history has always been fuelled by the synergy between revolutionary and extremist ideas and the types of communication prevalent then. This means that tools are always important for extremist forces in order to spread their ideas. Hate groups today have successfully brought about a war of ideas on the Internet. Thousands of radical sites exist in comparison to the very small number of sites that support moderate political ideas. Although physically a radical group may be located in a fixed geographical area, the Internet helps it to create a global impact. The advent of online media networks has enabled radical groups to find members and establish very real international networks through virtual means. Mehdi Masoor Biswas's Twitter handle @ShamiWitness is an apt example for the cross-border flow of radicalism. It is a fact that radical groups use Internet sites for activities such as recruiting members, broadcasting ideological statements or sermons, raising funds, and of course, inciting racial, ethnic, national and religious hatred.

It shows how far we have moved from the days of the radicals, handwritten pamphlets in a candle-lit basement to a more vibrant and articulate manner. While there are many factors that could indicate an individual committing on a terrorist act online, there is no smoking gun to determine who will actually act out. On the Internet, one can be anyone he/she wants to be. One can be Osama bin Laden, Hitler, Khomeini, or support their views in forums. However, an individual would not necessarily act that way in the real world. At what point does virtual and real life coincide? It is confusing to give an answer. Therefore, censoring is, in the long run, more confusing than an unrestrained freedom.

Even when there was no social media, no mobile phones and no television, the French Revolution took place, so did the Russian Revolution, and even in our country the Emergency was witnessed by hundreds. The spectre of the Emergency under Indira Gandhi is redolent. Rulers have to remember they cannot indulge in wrongdoings to curb people's aspirations. Prior to the Internet, we used ad pamphlets, town squares and loudspeakers to convey our ideas.

We have reached a point beyond which there is a catastrophe. Society will decay, we have every reason to think about the present role the Internet is playing in turning us into narcissists and spreading dissension and hatred in society, and it will continue unabated. However, censorship is not the panacea, it is just like firing the house for the fear of rats.

There are further difficulties associated with methods of monitoring, principally in the world of social media, which is ever more user-generated, interconnected and consisting of a wide range of content. Individual messages and emails are predominantly not easy to trace.

Instead of regulating the Internet, my personal belief is that social media use must be qualified, by which I mean the attempts that reduces the misuse of social media. In that direction, I suggest the following:

1. School and university curricula should be revised to sensitize the youth.
2. Children must be equipped with skills to verify fact and heresies in cyberspace.
3. Society should be well informed about distinction between gossip, rumour and hate and those of good and healthy information on the Internet.
4. Moreover, there should be strong filtering of content that spreads dissension and hate in the society. To that respect, help of intelligence organizations can be roped in.

That communally inflammatory attempts should not be tolerated on social media is a welcome thought, but the question is: is it doable? And is it possible to distinguish free speech from hate speech on the Internet? I guess, sometimes we have to pay a price for our freedom.

Yet, I believe that the people of a mature democracy should show more tolerance. Above all, what makes things right or wrong is our own judgement. If that is compromised, nothing can save the Internet from falling into the hands of miscreants for divisive activities. Although our ability to express ourselves and empathize with people has certainly been compromised, I hope we do not become mere slaves to technology.

Endnotes

1 PTI, 'Report Reveals Most Dangerous Online Scams in India', *Gadgets Now*, 10 March 2016, https://www.gadgetsnow.com/tech-news/Report-reveals-most-dangerous-online-scams-in-India/articleshow/51346131.cms, last accessed 24 May 2019.

2 Google, 'Transparency Report: Government Requests to Remove Content', http://www.google.com/transparencyreport/removals/

government/, last accessed 4 June 2014.
3. 'Internet Censorship: Law & policy around the world', *EFA*, 28 March 2002, https://www.efa.org.au/Issues/Censor/cens3.html, last accessed 4 May 2014.
4. 'Syria "Bans iPhones" Over Protest Footage', *BBC News*, 2 December 2011, https://www.bbc.co.uk/news/world-middle-east- 16009975, last accessed 20 May 2019.
5. 'After Blocking YouTube and WordPress, Turkey Now Bans Blogger', *The Road to the Horizon*, 26 October 2008, http://www.theroadtothehorizon.org/2008/10/news-after-blocking-youtube-and.html, last accessed 4 June 2014.
6. Elizabeth Flock, 'Google Is "A Spying Tool", Iran Police Chief Says', *Washington Post*, 10 January 2012, https://www.washingtonpost.com/, last accessed 8 June 2019.
7. Elizabeth Flock, 'Belarus Has Outlawed Browsing Foreign Websites', 3 January 2012, *Washington Post*, https://www.washingtonpost.com/.../belarus-has-outlawed.../gIQAhQNoYP_blog.html, last accessed 8 June 2019.
8. boingboing.net, 'Danish Police Proposal: Ban Anonymous Internet Use', 23 June 2011, https://boingboing.net/2011/06/23/danish-police-propos.html, last accessed 8 June 2019.
9. Yuliya Melnyk, 'Enemies of the Internet', *Reporters Without Borders Updates*, 16 March 2011, http://ijnet.org/blog/reporters-without-borders-updates-enemies-internet-list, last accessed 4 June 2014.
10. Zach Epstein, 'The Most Important Thing You'll See Today: Internet Censorship World Map', *BGR*, http://bgr.com/2014/02/20/internet-censorship-world-map/, last accessed 24 May 2019, Nick Routley, 'Map: Internet Censorship around the World', *Visual Capitalist*, 30 September 2017, https://www.visualcapitalist.com/internet-censorship-map/, last accessed 24 May 2019.
11. 'Names of Websites Blocked by Government Leaked Online', *Times of India*, 23 August 2012, http://articles.timesofindia.indiatimes.com, last accessed 5 December 2012.
12. Rama Lakshmi, 'India Blocks More Than 250 Websites for Inciting Hate, Panic', *Washington Post*, 20 August 2012, https://www.washingtonpost.com/world/india-blocks-more-than-250-web-sites-for-inciting-hate-panic/2012/08/20/aee0b846-eadf-11e1-866f-60a00f604425_story.

html?utm_term=.7cc46a2e06ee, last accessed 04 February 2013.
13. Anjana Pasricha, 'India Debates Misuse of Social Media', *VOA News*, 21 August 2012, http://www.voanews.com/content/india-debates-misues-of-social-media/1492129.html, last accessed 30 January 2013.
14. Jolie Lee, 'India Tops Countries Censoring Facebook Content', *USA Today*, 14 April 2014, https://www.usatoday.com/story/news/nation-now/2014/04/14/facebook-censor-india-turkey-pakistan/7694381/, last accessed 5 June 2014.
15. 'Freedom on the Net 2013: A Global Assessment of Internet and Digital Media', Freedom House 2013, http://www.freedomhouse.org/, last accessed 2019.
16. Nikhil Pahwa, 'DoT Issues Orders to Block 78 URLs; 73 URLs With IIPM Content', *Medianama*, 15 February 2013, http://www.medianama.com/2013/02/223-dot-block-iipm/, last accessed 5 June 2014.
17. TNN, 'Directed by Court, DoT Moves to Block 73 URLs Critical of IIPM', *Times of India*, 15 February 2013, https://timesofindia.indiatimes.com/india/Directed-by-court-DoT-moves-to-block-73-URLs-critical-of-IIPM/articleshow/18521319.cms?referral=PM, last accessed 24 May 2019.
18. IT Act 2000, http://eprocure.gov.in/cppp/sites/default/files/itact_contents/IT_DOC_NO_2/itact2000.pdf.
19. Department of Electronics and Information Technology: Information Technology Act.
20. Open Net Initiative, 'India', 2007, http://access.opennet.net/wp-content/uploads/2011/12/accesscontested-india.pdf.
21. Right to Information Act, 2005, Article 7(1), http://righttoinformation.gov.in/webactrti.htm.
22. Shivan Viv, 'Internet Censorship in India: An RTI Application', National Highway, 26 September 2006, http://shivamvij.com/2006/09/26/internet-censorship-in-india-an-rti-application.
23. https://en.wikipedia.org/wiki/Dalitstan.org
24. John Ribeiro, 'Orkut Comes under Fire in India', *InfoWorld*, 12 October 2006, http://www.infoworld.com/t/architecture/orkut-comes-under-fire-in-india-310.
25. Nikhil Pahwa, 'Our Views on Paid News in Digital Media & Blogs in India', *Medianama*, 21 June 2013, https://www.medianama.com/2013/06/223-paid-news-digital-media-medianama/.

26 'A Report by the Free Speech Collective', 31 December 2018, https://countercurrents.org/2019/01/free-speech-in-india-2018-the-state-rolls-on.

27 Rehman Azhar, 'Countering Hate Speech on Social Media', *Dawn*, 17 July 2012, http://dawn.com/2012/07/17/countering-hate-speech-on-social-media/, last accessed 12 June 2012.

28 Chris Gosnell, Hate Speech on the Internet: A Question of Context, *23 QUEEN'S L.J.* 1998, pp. 369–76

29 Natasha L. Minsker, '"I Have a Dream–Never Forget": When Rhetoric Becomes Law, a Comparison of the Jurisprudence of Race in Germany and the United States', *14 HARV. BLACKLETTER L.J*, 11.3, 1998, p. 113.

30 Richard Delegado and Jean Stefancic, *Must We Defend Nazis? Hate Speech, Pornography, and the New First Amendment*, New York: NYU Press, 1997.

31 Juliane Wetzel, 'Criminal Aspects for the Topic of Racism, Neo-Nazism and Right-Wing Extremism on the Internet', in Brigitte Bailer-Galanda et a l., eds., *The Net of Hate: Racist, Right-wing and Neo-Nazi Propaganda on the Internet*, 1997, available at law-wss-01.law.fsu.edu/journals/transnational/vol12_2/timofeeva.pdf.

32 Thomas Macaulay, Indian Penal Code, 1838, reprinting, 2002, p. 101.

33 Radio Poland, 'Poland to Create Internet Hate Speech Guidelines', 9 August 2012, http://eu.thenews.pl/1/9/Artykul/108671,Poland-to-create-internet-hate-speech-guidelines, last accessed 18 May 2019.

34 Calire Cain Miller, 'Google Has No Plans to Rethink Video Status', *New York Times*, 14 September 2012, https://www.nytimes.com/2012/09/15/world/middleeast/google-wont-rethink-anti-islam-videos-status.html, last accessed 18 May 2019.

Acknowledgements

Gayu, the love of my life, and my religion, I would not have done any of books, let alone this one, had it not been for you as a partner. I have developed all the ideas in this book through our conversations and your criticisms. I know you struggle a lot to adjust with the reality that I am preoccupied with my passion all the time. I know our time for conversations, travel, shopping, cooking and fun have all been compromised because of my serious preoccupation with writing. I know the battle you have fought in the silent chambers of your soul is all for me!

Nadha, for reading this manuscript on request and for which she made enormous efforts. She read all the whole manuscript carefully because of the love she has always had for me. I have learned a lot from her, and what this book is today is because of her. She told me that her entire perception about the Internet had changed after reading this book. Her comment touched me so much. But she read the draft three times! She is my undergrad student, who wishes to have her name on the cover of a book! Fara, for inspiring me a lot to write this book. She was also my undergrad student. Many ideas in this book took shape during our afternoon conversations. I owe a lot to her.

A generation of my students has deeply influenced me in writing this book. Special thanks to the generations of students

I taught in the beginning of my teaching career. I was able to relate with them so much and never felt the generation gap. Now, students who sit in front of me are those born of the iGen, who have developed the 'text neck' as they bend down their necks to look at their smartphones. Their 'text neck' culture made me curious. This shift in the student culture and change in the value system prompted me write this book. I owe to all of them. Thank you all!

Archana is one of the very few belonging to the iGen, who is still reluctant to download social media apps, even though she owns a smartphone. I don't know the reason behind this. She, too, has inspired me to write this book.

I cannot mention many due to privacy concerns, but I am indebted to a large number of people, hidden behind the fictitious names used in this book, who have contributed greatly to this book.

My friends, colleagues and relatives… connecting with them on social media has helped me develop my initial assumptions in the earlier stages of writing this book. I am indebted to all of you.

Special thanks to Yamini Chowdhury, senior commissioning editor at Rupa Publications, for her faith in my book. She is an outstanding publishing professional, who is able to come up with ideas within minutes. I owe her a lot. Priya Talwar, assistant commissioning editor at Rupa, who is very punctual with her assignments and took the painstaking job of developing the manuscript. Thank you, Priya. Special thanks to Anurupa Sen for copy editing the manuscript. To everyone at Rupa for the opportunity of sharing my ideas with a wider audience.

Acha, Ammachi and Shiju, for your love. All that I have done is just because of you.